SEDIMENT CONTROL AT INTAKES

CONTRIBUTORS:

P. Ackers	Binnie & Partners
P. Avery	BHRA
M.E. Bramley	Binnie & Partners - C.I.R.I.A.
P.A. Hammett	Sir Alexander Gibb & Partners
P.N. Hooper	James Williamson & Partners
P. Novak	University of Newcastle-upon-Tyne
A.R. O'Hea	Sir William Halcrow & Partners
K.V.H. Smith	University of Southampton
A.J.H. Winder	Watson Hawksley

The work in preparing this report was undertaken as part of Grant No. MEE67/0114, a contract within the Department of Trade and Industry's "Support for Innovation" scheme. The financial support from the Department of Trade and Industry is gratefully acknowledged.

SEDIMENT CONTROL AT INTAKES — A DESIGN GUIDE

Edited by P. Avery
(BHRA Process Engineering Division)

BHRA
THE FLUID ENGINEERING CENTRE

First published in 1989 by

BHRA, The Fluid Engineering Centre
Cranfield, Bedford MK43 0AJ, England.
Tel Bedford (0234) 750422 Telex 825059 Fax (0234) 750074

ISBN 0 947711 47 3

British Library Cataloguing in Publication Data
Sediment control at intakes.
1. Water pumps. Intakes. Design
I. Avery, P.
621.252

ISBN 0-947711-47-3

CONTENTS

Preface

This design guide is the result of a collaborative effort by a number of experts in the field of civil engineering hydraulics. The Department of Trade and Industry supported BHRA in setting up the panel of contributors, coordinating progress throughout the life of the project and editing the document.

The aim of the panel was to produce a guide which would enhance the engineer's understanding of the sediment problems associated with intake design, and the considerations involved and the options to afford a measure of sediment control at intakes.

The efforts of those involved are rewarded by the publication of this book, which is a unique compendium of hitherto uncollated advice and information concerning this subject. Of special interest to engineers working in the water, hydropower and agricultural engineering industries, we hope readers will find this guide helpful.

Pauline Avery, November 1988

NOMENCLATURE

A - mean plan area of settling basin

A_{gr} - value of F_{gr} at which motion first starts - related to efficiency of transport process

b - width of channel

C - local concentration of sediment (at elevation y above bed)

C_a - concentration of sediment (at reference elevation a above bed)

C_o - incoming concentration

C_r - concentration of suspended sediment removed

C_v - volumetric sediment concentration

d - water depth

D - particle diameter (D_{50} - median diameter)

D_{gr} - dimensionless grain size

eb - bed load efficiency factor

F_{gr} - sediment mobility number ($= \sqrt{Y}$)

Fr - Froude number ($= \frac{V}{\sqrt{gd}}$)

g - acceleration due to gravity

g_s - weight of solids passing/unit width/unit time

g_{si} - immersed weight of solid passing

G_{gr} - transport parameter

H - head over weir

I - dimensionless grain parameter

K - von Karman constant (0.4)

k_s - roughness (equivalent sand gradient diameter)

L - length of settling basin

m - performance parameter (m = 0 for 'best' basins; m = 1 for 'very poor' basins)

n - transition parameter (n = 0 for coarse sediments; n = 1 for fine sediments)

n_m - Mannings n

q - discharge intensity (volume/unit width/unit time)

q_t - sediment transport rate (submerged weight/unit width/unit time)

Q - discharge or flow rate

Q_c - canal flow rate (other subscripts defined in text)

r - radius of curvature (subscript i - inner wall; subscript o - outer wall)

Re - Reynolds number ($\frac{Vd}{\nu}$)

R - hydraulic radius of flow section (Area/Wetted Perimeter) (hydraulic mean depths)

s - specific gravity of solids

s_w - specific gravity of water

S - hydraulic gradient (or i)

SF - shape factor (= $a/\sqrt{bc}$ where a, b, c are mutally perpendicular dimensions, a being the smallest)

t_R - retention time

t_s - settling time

v_* - shear velocity (= $\sqrt{\tau_o/\rho}$ or $\sqrt{gdS}$)

V - mean velocity of flow

Vn - normal channel velocity

Vs50 - settling velocity of median size sediment

V_o - limiting velocity for zero sediment transport

V_s - flow velocity at incipient deposition

w - individual particle terminal velocity

W - width of settling basin

x - effective roughness coefficient

X - mass rate of sediment transport per unit width

Y - mobility number

Ycr - critical mobility number

z - Rouse number (= $w/\beta K v_*$)

Z - relative grain size (= d/D)

α - coefficient relating roughness, k_s, to median sediment diameter

$\tan\alpha$ - solid friction coefficient

β - ratio of sediment diffusion coefficient to momentum coefficient (assumed 1.0)

γ - unit or specific weight of solids

γ_s - submerged unit weight of solid phase

δ - reference laminar sublayer thickness

Δ - bed effective roughness

ε - sediment diffusion coefficient

η - sediment removal efficiency

λ - friction factor ($8gdS/V^2$)

ν - kinematic viscosity of the fluid

ρ - density or unit mass of fluid

ρ_s - density or unit mass of solids

τ_o - bed shear stress

τ_* - critical shear stress

ϕ - Einstein transport function

Chapter 1

THE PROBLEM OF SEDIMENT CONTROL AT INTAKES

1. THE PROBLEM OF SEDIMENT CONTROL AT INTAKES

1.1 Flow Diversion and Sediment Regime

The means of taking water from a river are numerous and varied. Whether for irrigation, water supply or hydropower the water being removed must pass through some man-made structure. No matter how slight, any interruption to the river course will cause changes to the river regime. The change being dealt with here is that to the sediment regime of the river. It is often the case that the proportion of sediment abstracted with relation to the total river sediment load is greater than the proportion of water abstracted.

The sediment load of a river may be classified broadly as bed load or suspended load. The bed load is made up of particles which move by sliding, rolling and saltation; the deciding factor being that bed load material properties are such that the particles cannot stay in suspension for long. The suspended load travels predominantly in suspension because the particle sizes and densitites are such that the turbulence in the flow does not allow them to settle out. The wash load is defined as those very fine particles which may only settle out in completely still water. The sediment transport mechanisms are fully described in Appendix 2, section A.2.2.

The nature of the transported sediment depends largely on the supply, from catchment erosion, landslips, etc., but also on the character of the river channel. Typically, near the river source, where velocities are highest, a large quantity of sediment may be carried by the river. This may be partly in the form of bed load, i.e. coarse sand, gravel, or larger stones and even boulders and partly as suspended load. As the river progresses to lower, flatter ground its velocity decreases and larger particles are deposited. As the river flows through sandy or silty lowland a greater proportion of the sediment transported will be of much smaller particle sizes, i.e. fine sand, silt and clay particles, and much of this will generally be in suspension. When the river is in flood the flow volume and velocity increase and with them the sediment carrying capacity of the river also increases.

As will be explained later (section 3.1), the bends in a river are areas of particular concern in terms of the process of erosion and deposition. This is due to the helicoidal pattern set up in the flow cross-section causing the bed load to be carried away from the outer bank and swept towards the inside, slower flowing area of the bend. This means that the water on the outer, or concave, side of a bend is clearer than that on the inside, a feature made much use of in the location and design of intakes.

Even when water diversion takes place via an intake structure on the outside of a bend, the curvature of diverted flow may develop in an opposite direction to that of the river if an inappropriate design is used, in effect creating an artificial bend flow with the intake on the inside of the bend.

Clearly, if a river carries sediment as bed load, suspended load, or both, it is impossible to abstract water devoid of sediment and, essentially, the hydraulics of flow diversion rule that the bed load will be drawn towards the intake structure or point of diversion.

So it is not surprising that numerous problems have been experienced world-wide with resulting high expenditure on remedial works or maintenance. This illustrates the need for careful planning in the design of an intake structure from the early collection of data, consideration of requirements, selection of site, etc., to the thoroughly studied design of the intake structure. The aim of this document is to enhance the understanding of the sediment problems associated with intake design, the considerations which must be taken into account and the options available to the designer to afford a measure of sediment control at the intake.

1.2 Problems Caused by Abstracted Sediment

If the problem of sediment ingress at an intake works is not considered, major difficulties will often result due to sediment being transported by the diverted flow, or by deposits caused by a reduction in the sediment carrying capacity of the diverted flow.

Small particles which remain in suspension, i.e. if the flow velocity is kept sufficiently high, may cause damage to any part of the intake works, particularly where machinery such as pumps (water supply) and turbines (hydropower) is involved. Exposure of such machinery to small abrasive particles moving through the impeller or turbine runner at high velocity causes damage over short periods of time; seals and bearings may suffer severe wear, so that efficiency is reduced and complete failure may eventually follow - in either case maintenance is necessary, requiring high expenditure in terms of replaced parts, man-hours and in lost water or power supply. Screens, penstocks and moving parts, such as gates and valves, are also susceptible to extensive damage from sediment in motion or from sediment deposits.

Wall, channel or pipe protection measures designed to resist exposure to water flow may not be sufficient to protect such surfaces against particle movement, and both cosmetic and structural damage may result.

In addition to the costs of damage due to sediment carried through intakes, there is the problem, particularly in the case of water supply, of removing the particles at treatment works, thus adding to the already expensive process of providing potable water to the consumer.

In cases where diverted flow velocities are not high enough to keep all particles in suspension - particularly in the gravity systems experienced mainly in irrigation schemes - the main problem is to avoid a greater intake of sediment than can be transported by the canal system. An excess of sediment will lead to deposition in the canal system, at first locally just downstream of the intake but gradually extending downstream to affect the whole canal system in time. The canal system will have strictly limited scope for adjusting slopes to accommodate the excess sediment supply, so freeboard will be lost as the bed rises and the capacity of the channel will decrease. The process of clearing and removing sediment is very expensive, especially if a proportion of fine cohesive material results in very solid deposits.

1.3 Examples of Problems Caused by Sediment

Bearing all these problems in mind, it should nevertheless be realised that it is seldom necessary to exclude all sediment from intakes and man-made waterways. This may be virtually impossible and would not be economically viable for many projects. Sediment in small quantities is often of value to land fed by irrigation systems. In cases where the abstracted water passes through machinery a small concentration of fine sediment is often permissible. But in the case of the high standards required for domestic water supply, considerable expense is involved in removing sediment and so the quantity allowed through the intake should be as small as possible.

The emphasis should be on CONTROL of the quantity of sediment which is permissible and this guide contains advice on methods of achieving this control. Control includes the exclusion or limitation of sediment entry and also the removal of sediment that gets through the intake. Once the designer is aware of how much sediment can be accepted, all possible design considerations should be made to ensure that this quantity is not exceeded.

The following examples are of intakes where serious problems arose because of the quantity or control of sediment.

(1) The Layong Intake, part of the Sungai Tutong Scheme in Brunei, (Fig. 1.1) had an original capacity of 0.75 m^3/s but this has been increased to 1.095 m^3/s. Raw water pumped to the treatment works was found to contain sand and silt.

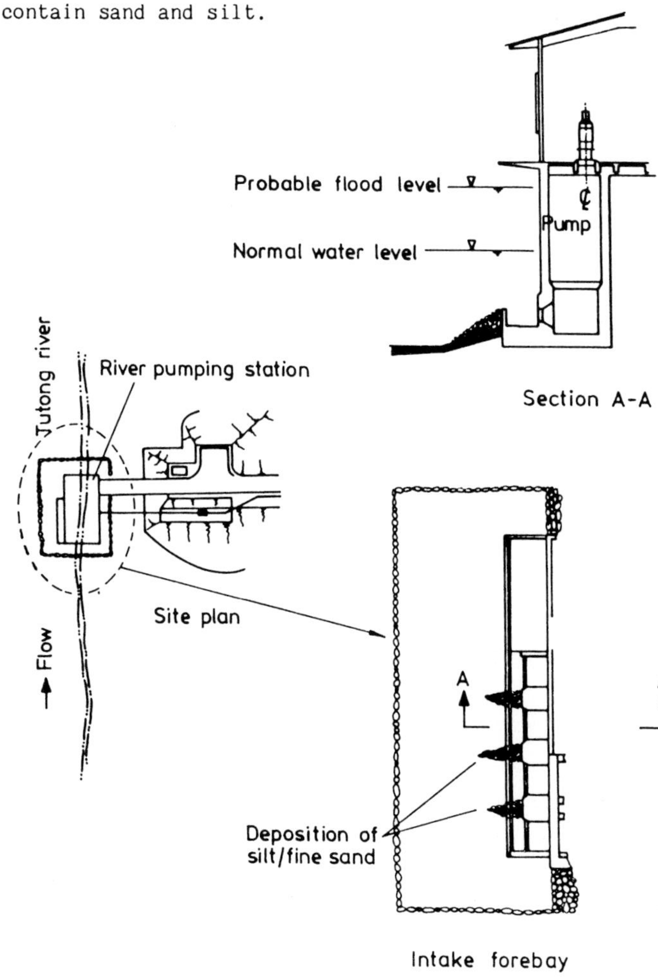

Fig. 1.1 Sungai Tutong scheme - River intake and pumping station

At low river level, investigation of the river bed immediately in front of the intake showed some deposits (0.3 m deep) of silt/fine sand on top of the stone pitching on both sides of the inlets to the pump sumps. The silt/sand deposits, which occurred directly in front of the inlets, are likely to have been drawn into the sumps.

Erosion of the pump impellers and casings were also found to be excessive due to the abrasive action of the sand and the low pH of the water. As a result, the running speed of the new pumps installed has been restricted to 735 rpm (from 1450 rpm), to diminish this.

Fig. 1.2 Cruachan aqueducts - Noe main intake and tunnel portal

(2) The Noe Intake, part of the Cruachan catchment area in Argyll, Scotland (Fig. 1.2) was commissioned in 1966. In 1977, a heavy thunderstorm in the vicinity caused approximately 1,000 tons of boulders to be washed into the intake choking it completely (Fig. 1.3). To remove the debris a tracked shovel was driven two miles through the Noe tunnel to the intake.

Fig. 1.3 Storm damage to Noe intake

(3) In 1973 in a paper entitled "Sediment problems of hydro-power plants". H R Sharma stated that the cost of abrasion repairs in one power house could be as much as $30,000 per annum.

(4) The turbines of the Florida Alta Plant in Chile (95 m head) were completely worn out after 2,000 hours of operation due to the presence of sand in the water.

(5) Muchamedon et al stated in 1975 that 50 million m^3 of sediment are removed from USSR irrigation canals every year.

Many additional examples where complete shutdown of plants is threatened because of intake blockages can be found in the literature.

1.4 Changes in River Regime

Records of the stage/discharge relationships of a river throughout the seasons and over as many years as possible are necessary before any river controlling or diverting (in part or in whole) structure is constructed (Chapter 2). These are essential to ensure that the required volumes and levels are available. In some cases a weir is constructed across the river to maintain and control levels - this is discussed in chapter 4 - but the design will nonetheless depend on good stage/discharge data.

With seasonal changes such as floods, snow melt-water, periods of drought, etc., the river discharge can vary with a ratio of maximum to minimum discharge of anything up to 1,000 or more, affecting sediment capacity and the sediment properties to a similar extent. The designer must study records of changes in river regime due to seasonal changes very carefully and should be aware of any other diversion, control or discharge structures planned for the stretch of river and its catchment under investigation, and take account of all these in his design calculations. Awareness of the life stage of the river - young, mature or old - is also important, so that trends already experienced or likely to occur can be considered.

Chapter 2

DATA REQUIRED FOR DESIGN

2. DATA REQUIRED FOR DESIGN

2.1 Purpose and Location of a River Intake

The purpose of a scheme and the distance of the river from the point of delivery of water will determine the length of the reach of the river on which the intake can be located economically. A pumped water supply scheme is less likely to be affected by topography than a gravity scheme in the same conditions. However, matters such as the required fall through the intake may be worth considering when the site for the structure is selected.

The site for an intake will be chosen on the basis of the local topography, geomorphology and on the geometry of the river channel close to the site. The requirements of the former subjects are dealt with in Section 2.2 below. The purpose of the system and the area selected for location of the intake are the first major items of data required for design.

2.2 Topographic and Geomorphologic Data

The initial selection of the site for an intake structure may often be made by study of available maps and/or aerial photographs. However, field inspection and local survey will be necessary in appraising the suitability of a particular site and in the design of training works and of the form of the entry to the intake.

Study of aerial photographs both current and previous issues, or failing that, inspection of the site on the ground or from the air will give some indication of the stability of the main river channel. Check should be made of evidence of previous channels in the river bed, any tendency of the channel to braid instead of maintain a single channel at low flows and the possibility of the river changing course within its flood plain. The presence of one or more old river channels does not necessarily mean that such changes of course have occurred recently and enquiry should be made of the local inhabitants as to the behaviour of the river. Protection of the existing river banks may be necessary to guard against the river adopting a previous course after a major flood; the stability and possible protection of the bank into which the intake is to be set must also be considered. If possible the river should be examined during floods.

Topographic surveys of the river channel will be required both upstream and downstream of the proposed intake site. The length of reach over which cross-sections are taken will depend on the initial assessment of the extent of hydraulic effects of the intake, river training works and bank protection works; particular attention must be given to accurate detail of the survey if hydraulic model tests are envisaged. Sections across the flood plain are often necessary where designs are required for rivers subject to peak floods in excess of bank full capacity. It is essential that the surveyors should also record the level of the natural water surface, location, date and time at least daily during the course of all survey work on the site.

A survey of the site of the intake structure itself should be carried out at a scale of 1:100 (or 1:200 for larger structures) with a contour interval of 0.5 metres or closer; the survey should extend onto the river bed. In precipitous country, larger contour intervals (say 2 m) may be adequate for large intakes.

Topographic and geomorphologic data which should be obtained, if available, would include:

- Existing reports and river studies.
- Topographic study of catchment and site - emphasis on features such as landslides and glaciers which affect stability of catchment.
- Aerial photographic study of catchment and site.
- Geological survey including records of river bed material (grading, mineralogy, etc.).
- Survey of soils and vegetation in catchment - and any likely future changes which may affect erosion.

Other features of the catchment which may influence the river morphology at the site, and therefore should be part of the survey, include:

- River control features (rock bars, gorges, etc.).
- Reservoirs and lakes upstream (existing and projected) which may act as silt traps.
- Trash producing industries (e.g. quarrying) and communities upstream.

2.3 Hydrometric Data

A knowledge of the flow at the site of the proposed intake is essential, and a stage-discharge relationship must be established.

Essential hydrometric data include:

- Magnitude of floods
- Flood hydrographs
- Frequency of floods
- Flow duration curve
- Stage-discharge curve
- Flow-sediment content relationship
- Average daily/weekly/monthly flows
- Meteorology - rainfall, wind speed, etc.

There are standardised measurement procedures available from literature (Ref. 2.1 (a)), including recommendations about the choice of site for obtaining good hydrometric data: often, the optimum site for an intake structure is not a good site for standard hydrometric measurements.

It is best to establish these hydrometric data by direct measurement using the normal methods of stream gauging, at a site close to the proposed intake location (so that no major tributary flows intervene). Where time is not available for gauge measurements due to urgency of design, an attempt can be made to calculate a stage-discharge curve on the basis of a survey of the shape and gradient of the river channel, the assumed effective roughness of the bed and open channel hydraulic formulae. Velocity measurements on a surveyed cross-section will provide some check on the assumed roughness of the bed. It must be recognised that this is an inadequate method of obtaining a good approximation to the stage-discharge curve at the site.

An assessment of the variation of river discharges and levels from year to year (including droughts) and with the seasons of the year, and correlation with the seasonal demands of the intake may be essential. (See Section 2.4 below). Where relevant this information will establish the probability of being unable to divert sufficient water into the intake to meet demands at low river flows. If this arises because of low levels rather than insufficient flow, provision of storage may have to be considered with construction of a low weir to raise minimum water levels and, the possible incorporation of gates in the weir to clear sediment from the forebay of the intake.

2.4 Operational Data

It is essential to establish the relationship between the discharge abstracted at the intake and the range of flows in the river through all the seasons of the year for effective design of an intake structure. The provisions of the design are generally to abstract the specified quantity of water of the required clarity from the source, however laden with sediment, and to keep the forebay of the structure free of deposits. The ability of the design to achieve this depends largely on the sediment carrying capacity of the main water course and the proportion of its flow which is being abstracted.

The pattern of intended operation of the intake throughout a typical year must be assessed so that the conditions at which the intake is at greatest risk from sediment ingress or from blockage due to shoaling in the river can be considered. The effects of large abstractions of water over periods of a few hours, as might be required where a reservoir is being supplied, must be taken into account. The resulting patterns of low and high discharges into the intake must be compared with the range of river flows and water levels at all periods of a typical year. This will enable the designer to be aware of occasions of high abstraction when the water supply is most heavily sediment laden and to make special arrangements to avoid or deal with these occasions. He will also be able to make operational arrangements for flushing and cleaning operations when the river is capable of disposing of deposits.

Account must also be taken of stages of development of the scheme which may require lower abstractions in its early years with much higher discharges planned for the same structure later in its lifetime.

2.5 Sediment Data

A full definition of sediment transport mechanisms is given in Appendix 2, section A2.2.

The bed load originates from the bed or banks of the river channel or its tributaries further upstream and the rate of transport is dependent on the velocity and turbulence of flow at any time. Reliable measurement of bed load is very difficult and it is common practice to derive an approximation by the methods described in section 2.6 below. Some assessment of the order or magnitude of the movement of the bed of the river is required in the choice of elevation of a sill at the entry to the intake structure to exclude the bed material, and in the provision to be made at any cross-river weir to deal with deposition behind the weir. Clearly, abstraction of the near-bed flow which contains these high concentrations of sediment should be avoided.

The suspended sediment may originate from the same source as the bed load or it may be the wash load, i.e. much finer material washed into the river by heavy rainfalls eroding the catchment. This material will be drawn into the intake and will either remain in suspension or settle out slowly in relatively still conditions. The concentration of suspended solids at a range of discharges of the river must be determined; knowledge of the gradings of these materials is also necessary to the design. Methods of sampling the suspended sediment are described in Section 2.6 below.

These samples must provide a flow-sediment content relationship (Section 2.3) for the supply to cover all seasonal variations.

The recorded measurements of sediment concentration in the river adjacent to the proposed site of the intake may be assembled to give a series of curves of sediment concentration against the reduced levels of the points of sampling. The resulting curves for various flows at the time of measurement will indicate the quantities of sediment liable to enter the intake for a proposed elevation of the intake sill. The result of such an assembly of data is illustrated in Fig. 2.1.

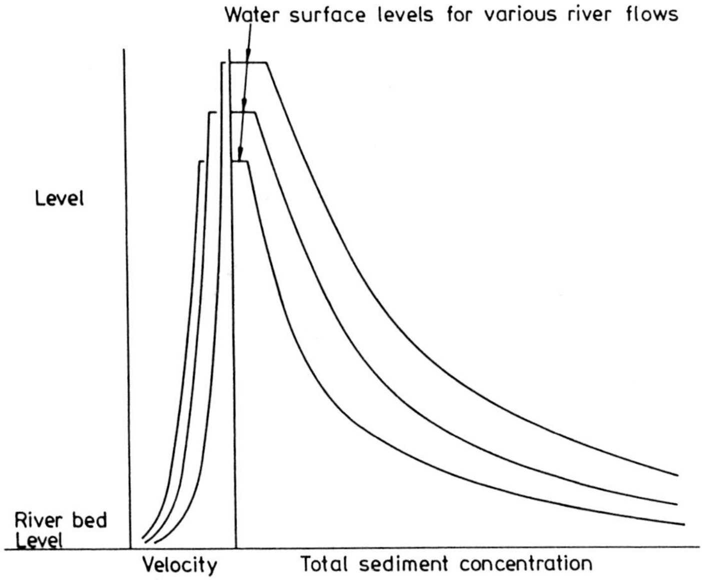

Fig. 2.1 Distribution of suspended sediment and velocity

The grading of the sediment at each level can also be informative as to the problems of later disposal of that part of the river sediment drawn through the intake. A typical presentation of comprehensive data for a single river discharge is illustrated in Fig. 2.2.

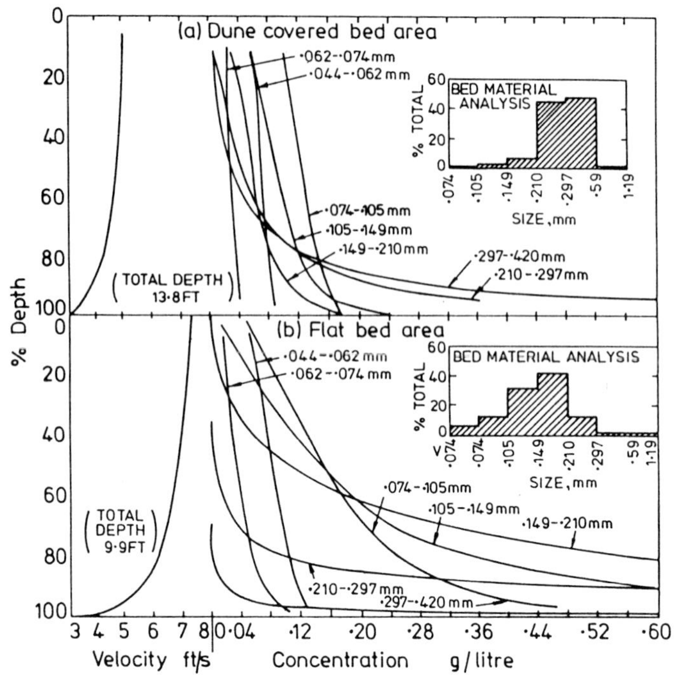

Fig. 2.2 Distribution of suspended sediment and velocity for Missouri river at Omaha

The constuction of an intake structure abstracting a high proportion of the supply will of course modify the pre-construction sediment concentration and distribution. So it is very important to consider the effects the intake structure and the flow abstracted by it will have on the source. The ability of the river to dispose of flushed or sluiced sediment and the effects that this will have downstream must also be considered.

2.6 Data Collection Techniques

2.6.1 Collection of Hydrometric Data

The procedures and equipment required to record hydrometric data are readily available in literature such as Ref. 2.1. The points given below are applicable to most situations.

Where it is necessary to measure directly the flow in the river to calibrate the level gauge, a cross-section should be chosen where the river reach is fairly straight with a stable channel and with streamlined flow under most stages of river discharge. The velocity of flow is taken at a number of depths on a vertical line on the cross-section, the depth of water at that location being recorded, and preferably at equal horizontal intervals across the river. The set of readings must be repeated for various water levels to obtain a level/discharge rating curve. A propeller actuated meter is commonly used but there are other types of meter that may be appropriate - cup meter, electromagnetic meter, drag meter, pitot tubes, etc.. To carry out gauging during periods of high river flow, wading through the river with a velocity meter becomes impracticable and recourse has to be made to an anchored boat or to the erection of an overhead cableway from which to take measurements.

2.6.2 Bed Load Data

Other than the case where a reservoir downstream is trapping the sediment carried by the river, only an approximation to the actual bed load of the river can be obtained. There are two methods of doing this. The first is to attempt to measure directly the entry of material into a trap placed on the river bed, but, although a number of pieces of equipment have been used (see Fig. 3.26 of Ref. 2.2), results are of uncertain reliability. The average of many measurements is required because of a high degree of variability in bed load transport. The second method is to deduce the bed load as a proportion of the suspended sediment load (see Table 3.2 of Ref. 2.2).

2.6.3 Suspended Sediment Data

An adequate estimate of the suspended sediment load can only be obtained by taking samples on a regular pattern across the river and at a range of depths; the sampling must then be repeated for various discharges of the river. Suspended sediment concentration may not only vary with changing river discharge but also for any given discharge because of other factors. This variability should be identified. A number of forms of sampling equipment are in use which are illustrated in more detail in other publications (see Figs. 3.11 to 3.17 of Ref. 2.2) but the main points of the devices are described below. All these sampling instruments must be positioned in the stream so that the following requirements are met: the flow velocity at the intake to the sampler must be representative of the velocity in the part of the cross-section of the river at which measurements are being taken; the disturbance to the streamlines of flow must be minimal; the intake of the sampler must be correctly orientated vertically and horizontally. Depending on the size of the river, the operator may be wading, in a boat or on a cableway.

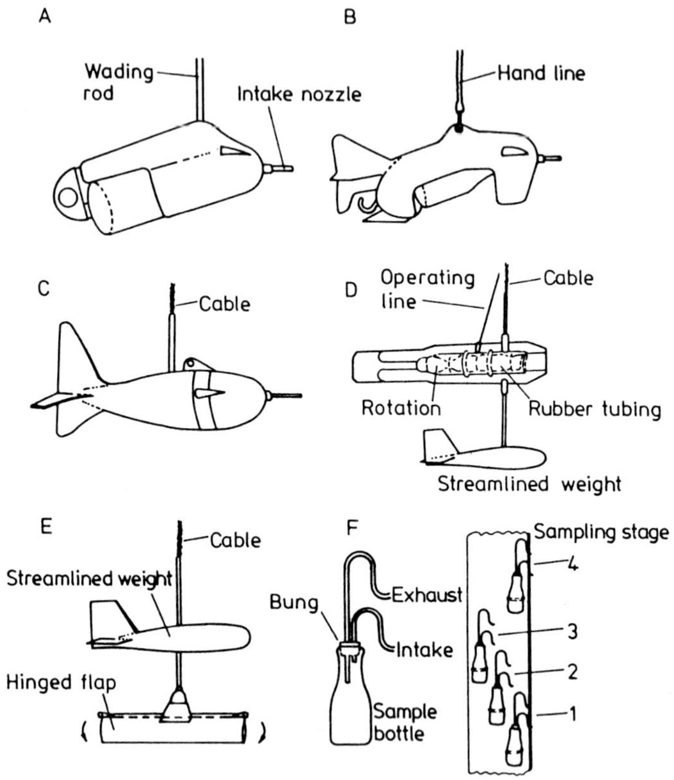

Fig. 2.3 Suspended sediment sampling equipment

(i) <u>Instantaneous Sampler</u>

This is the simplest type but it makes no allowance for turbulent fluctuations of concentration. It consists of a horizontal tube open at both ends which is lowered to the required depth and then the ends are closed (see D and E of Fig. 2.3).

(ii) Time-Integrating Point Sampler

The sampler illustrated at C on Fig. 2.3 overcomes the problem of fluctuations of concentration by filling over a short but significant time interval but is representative of that point in the cross-section only. Samples are collected at selected depths at stream verticals representing areas of equal water discharge in the cross-section.

(iii) Depth-Integrating Sampler

The sampler (illustrated at A and B on Fig. 2.3) is lowered to the stream bed and then raised to the surface at a constant rate. The resulting sample is a discharge-weighted mean concentration for the vertical location. The rate of raising of the device must be chosen by trial and error so that the sampler is not completely filled on its return to the surface. The verticals across the river are selected as in (ii) above; the rate of raising on a vertical must be constant but need not be the same as on other verticals.

(iv) Single Stage Sampler

Where a sample is required at high flows a simple unmanned type of sampler is illustrated at F on Fig. 2.3. The bottle is mounted at a predetermined location and level (above the current water surface) and a sample is taken slowly on the rising stage and can be collected later.

(v) Pumping Samplers

A refinement of the automatic sampler is a programmed device to abstract samples at pre-set time intervals or to be activated when the water level reaches a certain stage.

The concentration of suspended sediment and the velocity of flow vary with the cross-section and the sampling technique to be adopted should enable the total load to be estimated with the minimum of samples.

The data normally obtained from a set of sediment samples would include:

- distribution of suspended sediment load across the river section
- particle size and grading curves
- specific gravity of particles

In addition the site data obtained would also include:

- temperature of the river water
- discharge of the river and the distribution of velocity across the section

Use of the sediment data gathered by the techniques outlined above is described in Appendix 2, section A 2.9.

References

2.1 Chapters (a) A8 and (b) C2. In Applications in Hydraulics, Book 3, Techniques in Water Resources Investigations of the US Geological Survey. US Govt., Printing Office, 1970.

2.2 "Sedimentation Engineering" - Editor Vito A. Vanoni - published by the American Society of Civil Engineers (ASCE manuals and reports on engineering practice - No 54), 1975.

2.3 "Drainage Basin Form and Process" - Gregory, K.J. and Walling, D.E. London : Edward Arnold, 1973.

Chapter 3

INTAKE LOCATION AND ALIGNMENT

3. INTAKE LOCATION AND ALIGNMENT

The choice of site for an intake is governed by many factors which are outlined in Chapter 5. However, particularly with regard to river intakes, there are useful guidelines for the location and alignment of an intake structure which enhance the ability to control sediment ingress to the intake.

3.1 River Bend Hydraulics

Close to its source a river is generally of less capacity and higher velocity and is less accessible than when it reaches the flatter lowland regions. It is in the former regions that a river is most susceptible to surrounding geology and will wend its way via the easiest route to the sea, lake or other river into which it discharges. In the lowland areas bends are major features of rivers. When considering a site for an intake the effects of bends on the velocity distribution across the river section are very important.

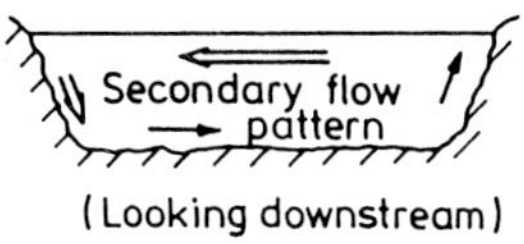

Fig. 3.1(a) Secondary motion generated in a bend to the right

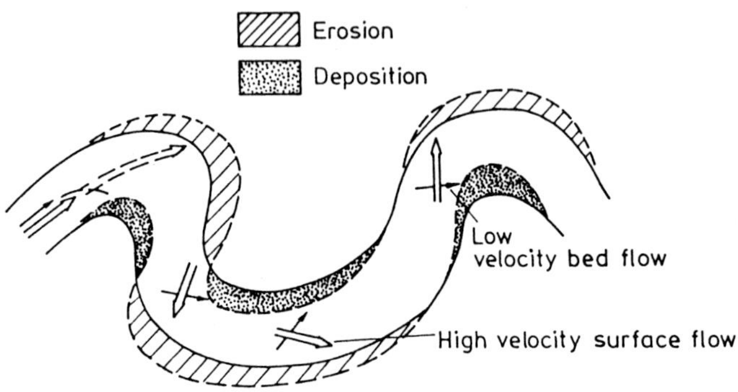

Fig. 3.1(b) Erosion of the outer bank of a bend caused by the faster moving component of the stream, and deposition of silt on the inside where the slow moving bottom water rises

The cross gradient due to the bend causes super-elevation which sets up secondary flow as shown in Fig. 3.1. The spiral flow current carries the less heavily sediment laden surface flow to the outer or concave side of the channel and the more heavily laden lower layer of flow together with any bed material in movement, to the inner or convex side.

This indicates that near a bend the outer bank is the appropriate position for siting an intake where sediment ingress is to be avoided. Further, the effect of sediment exclusion is known to be more pronounced when the intake is located towards the downstream end of the bend where secondary currents have become fully effective. The river should be well established and the banks stable in the region where the intake structure is proposed. If most of the sediment in the river is carried in suspension, this method of avoiding sediment ingress is not so effective as where most movement is in the form of bed load, but it nevertheless will encourage the removal of cleaner water.

3.2 Alignment of Intake

The intake should be aligned to the main flow to produce a suitable curvature of flow into the intake. This means essentially that the flow direction should be changed as little as possible. In this way the flow will behave as though the intake was on the outside of a curve and so bed load will be swept away from the vicinity. If, on the other hand, the flow is diverted by a large angle the flow patterns will be disturbed and bed load will be attracted towards the intake.

References reviewed in 1981 (Ref. 3.1) recommended diversion angles of between 10° and 45°, i.e. the angle between the intake centre line and the main direction of river flow. The ideal angle for a particular intake depends on the ratio of abstracted to river flow, the widths of the river and intake forebay and other factors. It is fair to say that a diversion angle of less than 45° is favoured and the optimum angle could be achieved by observing flow patterns in an hydraulic model.

3.3 Hydraulics of 90° Diversion Angle

It is repeatedly found that despite publications with recommendations to the contrary, intakes are often constructed in river banks with the intake axis perpendicular to the river bank. This means that the flow has to be diverted through 90° to enter the intake structure. The associated problems of flow separation and recirculation and sediment deposition often occur. If no attempt is made to improve the curvature of the flow approaching the intake these problems are insoluble. Methods of improving the flow curvature are included in later chapters. The drawing in Fig. 3.2 illustrates how the near-bed flow is attracted towards a 90° intake. The arrows indicate the direction of the near bed flow and S_1 and S_2 the stagnation points. Guide vanes or walls can improve the flow curvature into the intake.

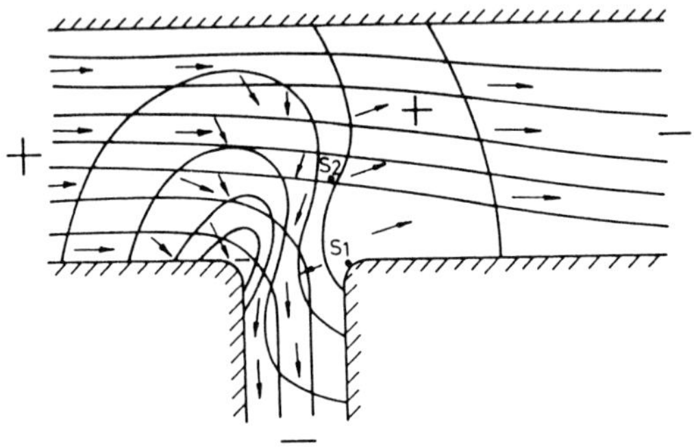

Fig. 3.2 90° Offtake

Examples of problems caused by bad siting of an intake often appear in literature. There are instances where complete re-location or re-construction of the intake has been carried out. The location and alignment of the intake are the earliest most important considerations in its design.

References

3.1 Avery, P. Sediment Exclusion at Intakes - A Review, BHRA Report RR 1725, August 1981.

Chapter 4

RANGE OF INTAKE STRUCTURES

4. RANGE OF INTAKE STRUCTURES

4.1 Types of Intake Structure

The prime purpose of an intake is to allow abstraction of water from the source with as little sediment as possible, thereby minimising maintenance and operational costs, and providing some measure of protection against damage to, or blocking of, the conduit by incoming sediment. A number of distinct types of intake structure have been developed, the selection of which is likely to depend on the location, scale and function of the project.

Intakes can be from a water course, reservoir or sea. The main emphasis in this guide is on water course intakes where sediment control is often a major design constraint.

Typical features of intake structures are described in Appendix 1.

4.1.1 Bank Intakes

These are structures located on a river or canal bank, side of reservoir or a coastal site. They are generally adopted for locations where only a small portion of the flow passing the intake is to be abstracted and where fluctuations in water level are not large. Bank intakes are appropriate for irrigation, water supply and power functions.

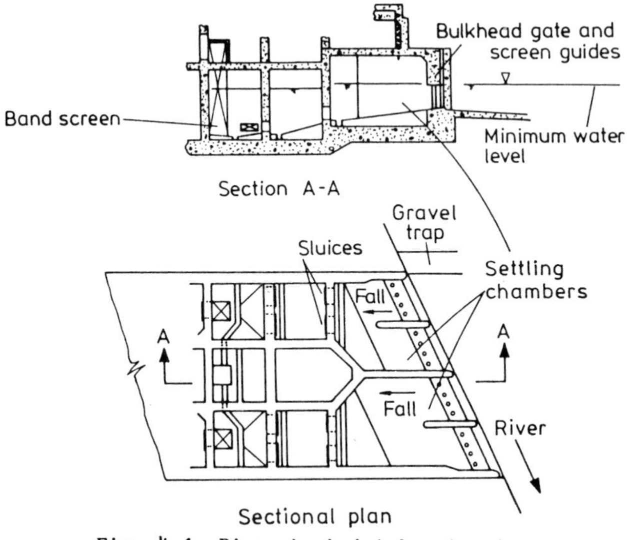

Fig. 4.1 River bank intake structure

Fig. 4.1 shows a typical river bank intake for a water supply pumping station, where sediment transport in the river is normally not significant. The face of the intake is aligned with the bank. The intakes, which are at bed level, have coarse screens, bulkhead gates and fish electrodes. Behind the coarse screens there are settling chambers to trap coarse material that might enter, and band screens and finally the pump chambers. The maximum abstraction rate is 350 Mℓ/day. The bed material in the river was composed of approximately five equal fractions of the following size ranges: 175-75 mm, 75-33 mm, 33-17 mm, 17-5 mm and below 5 mm. A gravel trap across the river just upstream of the intake was found through model studies to be efficient in protecting the intake from the ingress of bed load.

4.1.2 Side Intakes with Cross Weirs

For rivers and streams where a substantial proportion of the flow is to be diverted, a cross weir of some sort is an essential feature to ensure available water is not lost to the intake at low stages.

Fig. 4.2 shows a layout for a side intake with cross weir, supplying a free flowing pipeline. The arrangements work well provided solid and floating debris are not present in large quantities, and if screen cleaning can be undertaken regularly. The average stream flow is 50 ls^{-1} and the design intake flow (typically five times the average) is 250 ls^{-1}. There is about 10 m^3 of sediment storage available in the head pond, which would require shovelling to clear, this in turn may require suitable access arrangements. Sediment sizes encountered are in the range of coarse sand to gravel plus a few cobbles.

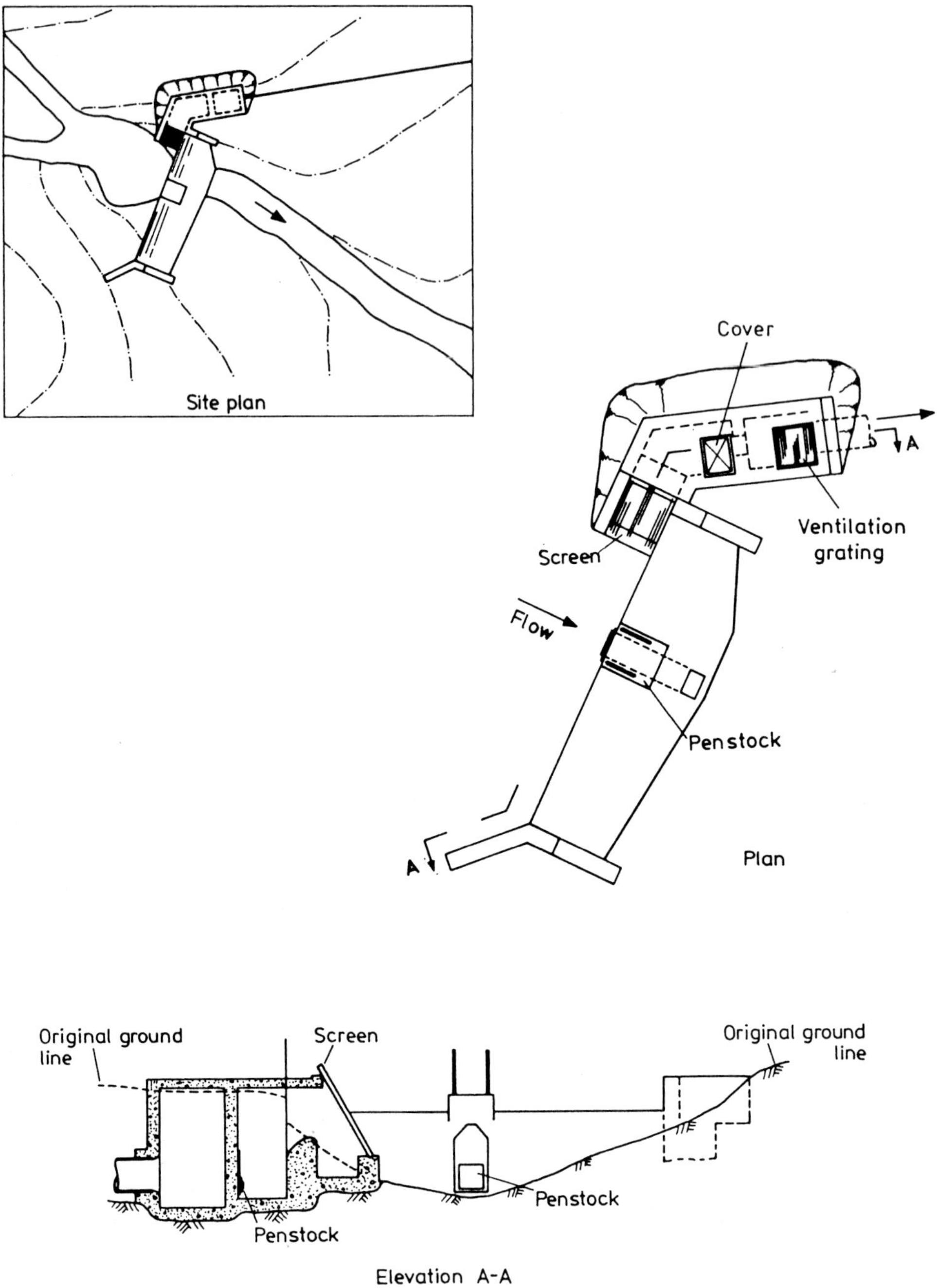

Fig. 4.2 Side intake with cross weir

A modified version of the side intake with cross weir is shown in Fig. 4.3. The screens have been eliminated, and a skimmer wall excludes floating debris. Typical flow rates here are an average stream flow of 110 ls^{-1} and design intake of 1000 ls^{-1}. The sediment sizes are in the coarse sand and gravel range.

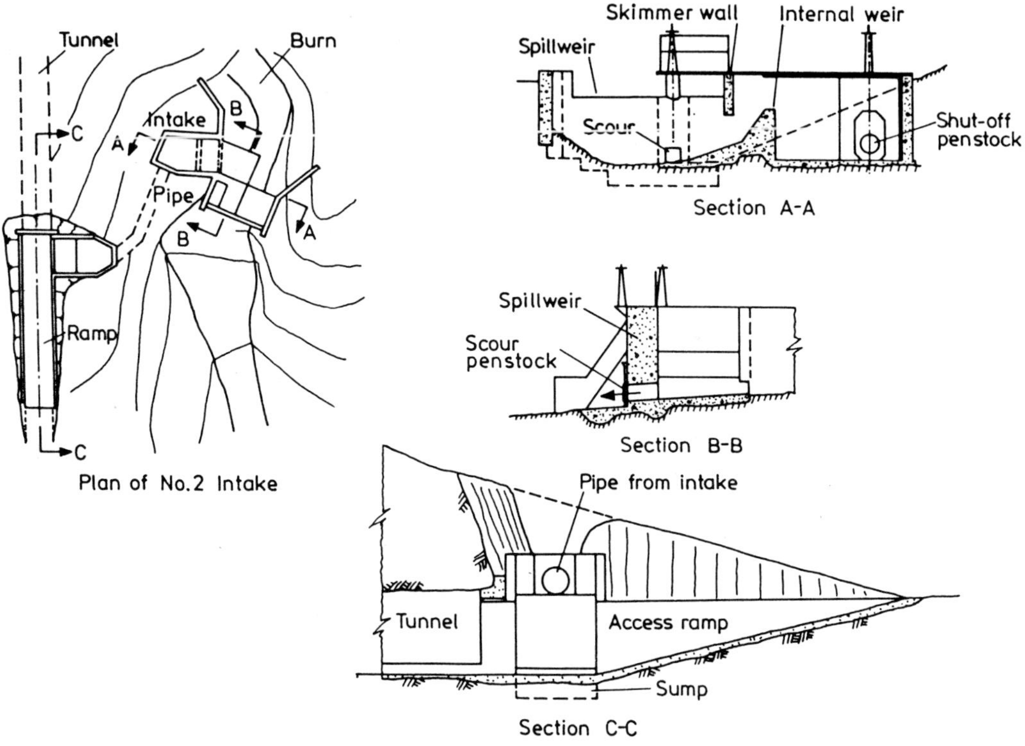

Fig. 4.3 Screenless side intake

Where a heavy sediment load is carried along in suspension or near the river bed additional arrangements may be necessary to minimise sediment influx and to avoid blockage of the intake by shoaling in the river. Fig. 4.4 shows such an arrangement, typically for an irrigation or power canal intake (Ref. 4.1). The function of the main features of the intake are described in Appendix 1. In this case the desilting canal is used for intermittent return of desilting flow to the river.

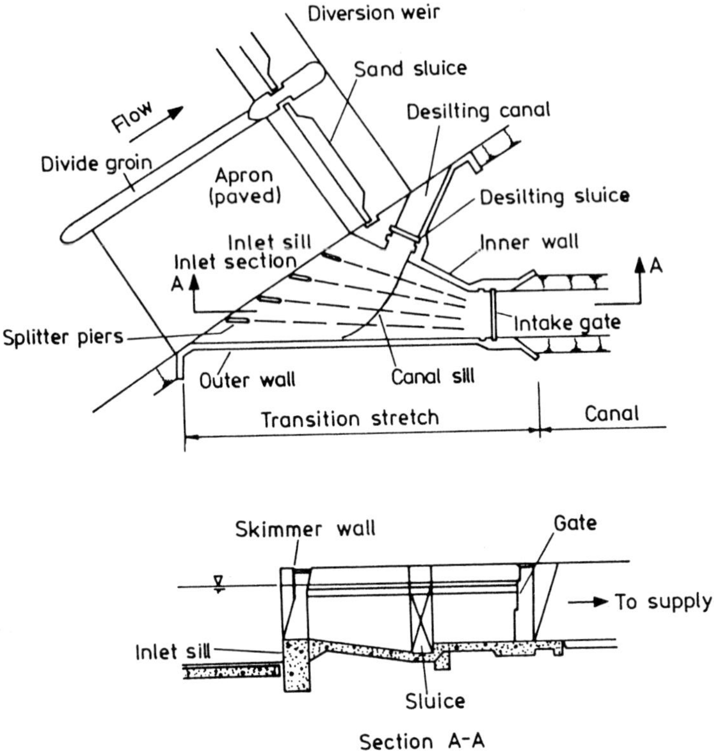

Fig. 4.4 Side intake on river with heavy bed load

4.1.3 Low Head River or Canal Diversion Works

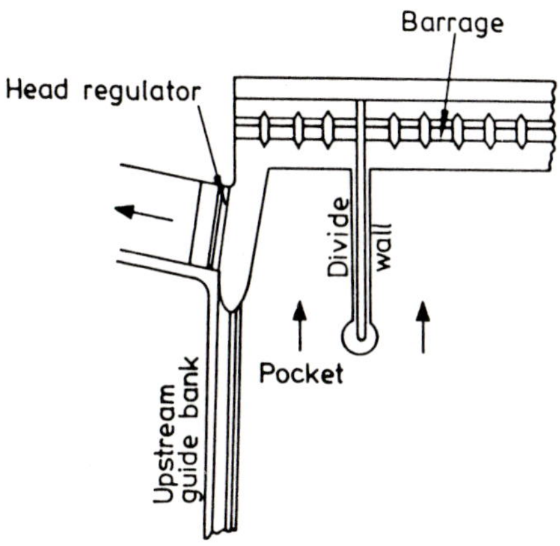

Fig. 4.5 Typical canal diversion works

As previously mentioned weirs are constructed across rivers at intakes to maintain adequate levels and volumes of water for abstraction. For irrigation systems barrages are often constructed across main water courses at the headworks of new canals. The barrage and associated canal headworks are designed to divert water from the main water course, to control sediment entry into the irrigation canal system and to control the distribution of water within the canal system. Fig. 4.5 shows a typical canal diversion works.

Some situations may require continuous sluicing of bed load, and a successful arrangement is shown in Fig. 4.6 (Ref. 4.1). Here the intake to a power canal has a series of undersluices set in the face of the intake sill, which extract the lowest layer of flow with the highest concentration and coarsest sediment, and return it continuously to downstream of the river control structure.

Further examples of the methods used to control sediment entry to canals at diversion works are described in Chapter 6.

4.1.4 Bottom Intakes

Bottom intakes have been developed for glaciers and mountain torrents, where site conditions may be extremely difficult for access and construction, and where boulders and rock debris have to be passed with minimum obstruction.

Fig. 4.7 shows the arrangement for a 'Tyrolean' type of bottom intake for a hydro power scheme in the French Alps (Ref. 4.2). It comprises a collecting chamber across the bed of the stream covered by a coarse screen. The total stream flow passes over the chamber, and the screen admits fine

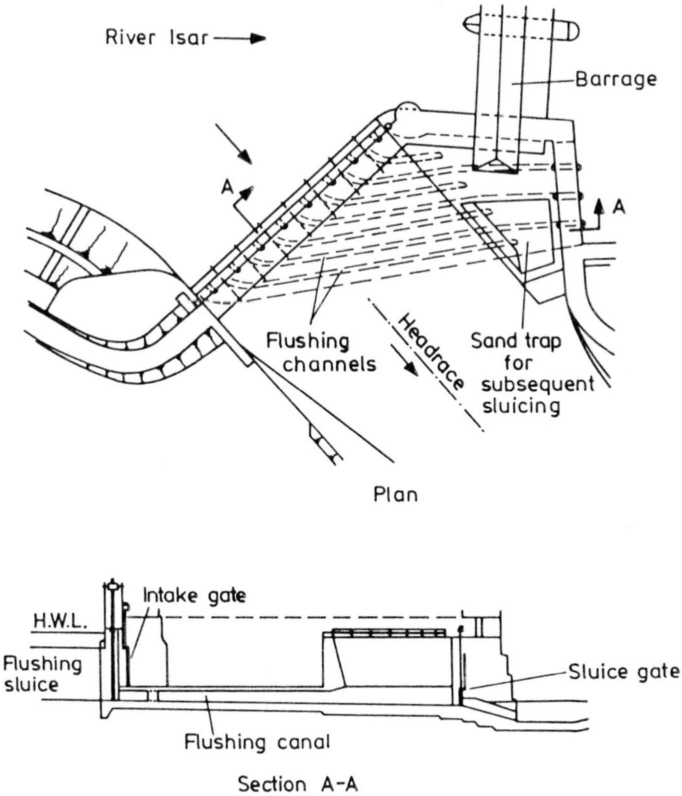

Fig. 4.6 Side intake with undersluices

debris with the water entering the chamber. Excess inflow is spilled at the downstream sill. The conduit from the collecting chamber is designed for debris which has entered to be carried with the flow to a settling basin constructed a short distance downstream. Here clean water is skimmed off for the power scheme, and arrangements made for periodic scouring of settled debris. Typical intercepted discharge is 3 m^3 s^{-1}.

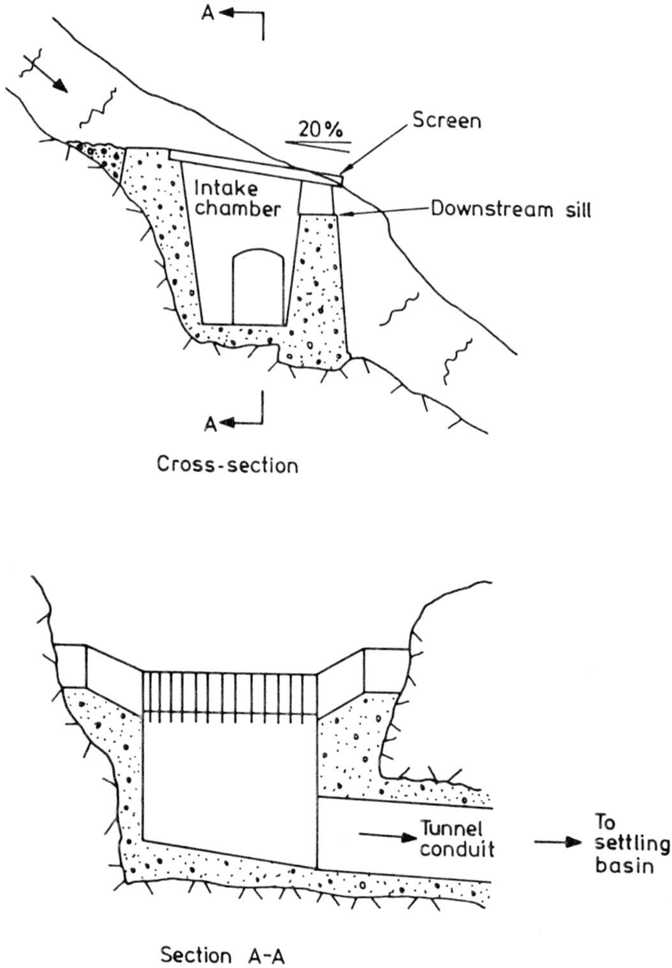

Fig. 4.7 Tyrolean intake

Smaller versions of bottom intake are shown in Figs. 4.8 and 4.9. These designs were developed for diversion of side streams to hydro power reservoirs. Steeply sloping screens with round bars were found beneficial in minimising blockage and loss of water. Light weight debris may be carried through the aqueduct to the reservoir; heavier material trapped in the screen chamber, is scoured by a hand operated valve. The maximum design intake flow of these examples are 150 ls^{-1} and 750 ls^{-1} respectively.

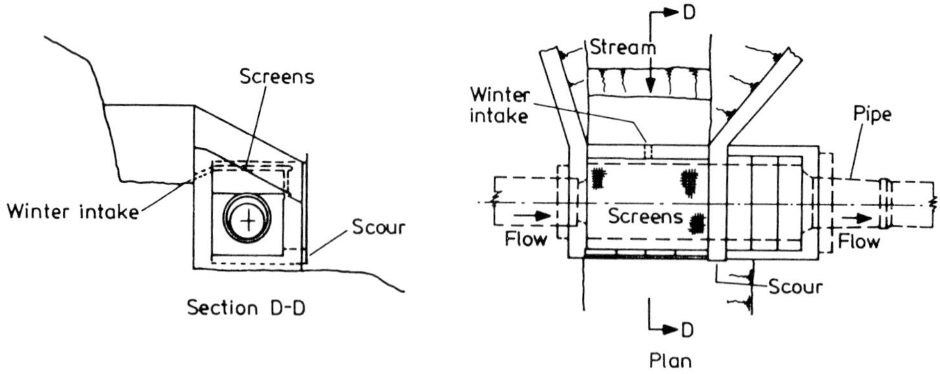

Fig. 4.8 Pipeline bottom intake

Fig. 4.9 Photograph of bottom intake to tunnel

4.1.5 Frontal Intakes

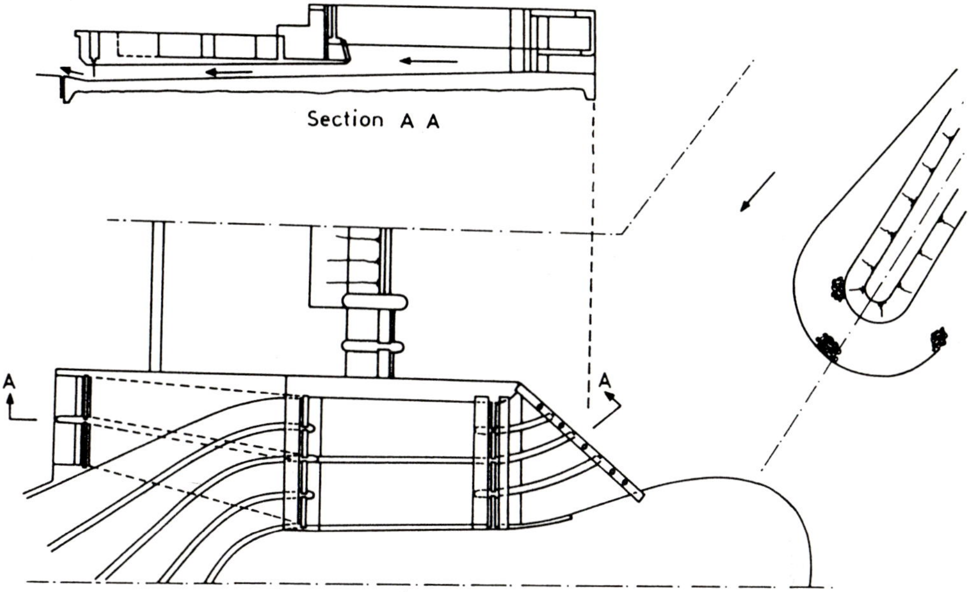

Fig. 4.10 Design of frontal intake (from Cecen - see Ref. 3.1)

A frontal intake designed for abstraction of clear water from mountain streams is shown in Fig. 4.10. This design has been used successfully in major systems mainly in Turkey. Since the abstracted water is taken from the upper layers of the river while the lower layer is continually flushed past the intake, it is particularly applicable where the majority of sediment carried by the river is bed load, and where a large proportion of the flow continues down the original water course.

4.1.6 Submerged Intakes

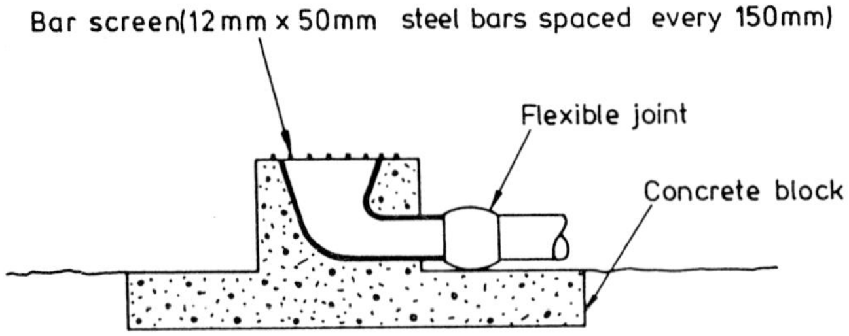

Fig. 4.11 Submerged intake

A simple submerged intake is shown in Fig. 4.11 (Ref. 4.3). It comprises a bellmouth and bend set in a block of concrete on the bed of a river or reservoir, with connecting draw-off pipe. The inlet is protected with a bar screen and set high enough above the bed to allow sediments to pass on either side of it if used in a river. Such an arrangement may be used for the drainage outlet of a small reservoir.

A more elaborate concept with vertical shaft, for larger flows is shown in Fig. 4.12. A framed structure extends to above top water level, allowing access for maintenance of the screens and shaft. The structure was planned as a cooling water intake in a shallow estuary. Here the screen sills must be set above the highest level to which estuarial sediments might be expected to rise. Each of two intakes here provides 36 m^3s^{-1} of cooling water.

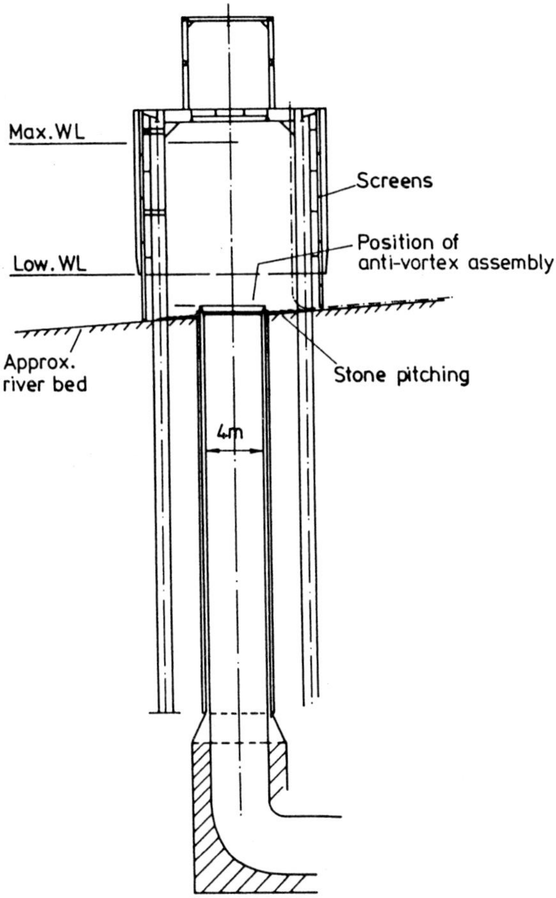

Fig. 4.12 Submerged shaft intake

Both intakes (Figs. 4.11 and 4.12) are located nearly at bed level and allow a maximum range of water levels to be utilised.

4.1.7 Tower Intakes

These are generally utilised where there is a large water level variation such as in storage reservoirs or tidal waters, and access for operation of gates or valves is essential. They are free standing structures, set in deep water, usually with an access gantry above top water level connecting to the shore.

Some tower intakes are "wet" in that water fills the inside of the tower to approximately the same level as outside, under normal operating conditions. The towers are dry only when the gates are closed and the interior dewatered. A "dry intake tower" is illustrated in Fig. 4.13. The tower houses a vertical withdrawal conduit arranged with individually gated branch inlets, at several different levels, allowing draw-off from any selected level. The bottom outlet valve is designed to scour out any sediment collecting adjacent to the intake.

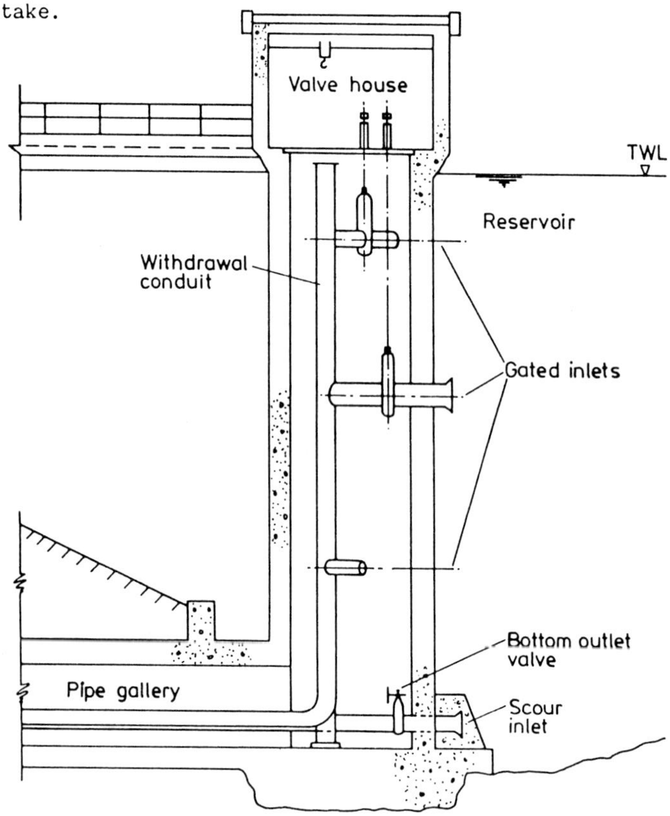

Fig. 4.13 Dry intake tower for water supply

4.2 Choice of Intake

In selecting a particular type of intake where there is a requirement for the exclusion of sediment, factors to be considered should include the function of the intake, the scale of the works in terms of flow quantity and range of depth, particular features of the site and method of construction.

Principal functions for intakes include water supply, power generation, cooling water, irrigation, stream diversion and drainage. The types of intakes shown in Figs. 4.1 to 4.13 indicate generally the function appropriate to the type of intake, and some of the variations that may be adopted for the exclusion of sediment. Chapter 6 outlines further devices which may be incorporated into intake designs to exclude and control sediment.

The data collected prior to site selection (as described in Chapter 2) and the sizing of the intake (Chapter 5) will provide limitations to the choice of intake structure type. Knowledge of the nature of the river, its range of levels, discharges and sediment load, will indicate whether the intake should be a bank or bed intake for example. The required operating criteria of the intake structure and associated distribution works provide further restrictions to choice. Whether or not there will be sufficient water to spare for flushing and sluicing operations will affect the choice of methods for dealing with excess sediment.

Where an intake is to be constructed for a large flow, such as for power generation or cooling water, the intake location and hence its type, is most likely to be determined by the overall layout adopted for the hydraulic system. In shallow water such as an estuary, a submerged vertical intake may be the only possible choice, and in such a situation, a particularly careful investigation of the sedimentation of the area would be an essential prerequisite.

A new intake site may be on a stream, river, canal, lake, new or existing reservoir, or coastal location. Access to the site, and whether construction is to be in the dry, in cofferdam or through water, will all be considerations in assessing the most appropriate type of intake.

In all cases model studies provide valuable information on the performance and suitability of the intake structure design.

References

4.1 Mosonyi, E. Water Power Development, Vol. 1, Hungarian Academy of Sciences, Budapest, 1957.

4.2 Ract-Madouse, X. et al. Some recently built high altitude bottom intakes, La Houille Blanche, N. 6, Dec. 1955, p.853.

4.3 Linsley, R.K. and Franzini, J.B. Water-Resources Engineering, 3rd Edition, McGraw-Hill Kogakusha Ltd., Tokyo 1979.

Chapter 5

DESIGN OF INTAKES

5. DESIGN OF INTAKES

This chapter incorporates general considerations for intake design in addition to those specific to sediment control.

5.1 Design Procedure

5.1.1 Requirements

The basic requirements for the supply of water must be defined:

(i) Maximum and minimum abstraction rates during the different seasons of the year.

(ii) Maximum sediment content and the maximum grain size of particles that can be accepted into the supply system.

(iii) The acceptability of partial or total failure of the supply and the corresponding return period in years.

(iv) The location and relative level of the point of delivery of water from the intake must be defined so that a choice is made between gravity and pumped delivery and also between open channel and pressure pipe conveyance of the water.

(v) Provision for future expansion of the demand.

5.1.2 Limitations

The above requirements must then be compared with the limitations of the source of water:

(i) Maximum and minimum water levels in the river or reservoir at the accepted probability of restricted supply (see 5.1.1 (iii) and (iv) above).

(ii) The corresponding maximum and minimum flows available.

(iii) The concentration of sediment at these flows and the reduction in concentration as the level of the entry sill of the offtake is raised above the bed (see Section 2.5 and Figs. 2.2 and 2.3).

The effects that the abstraction will have on the sediment concentrations and distribution must also be considered.

5.1.3 Location of Intake

The considerations limiting the location and alignment of the intake are described in Chapter 3 above.

5.1.4 Type of Intake

After consideration of the requirements of the intake and of the limitations of the supply, a choice of the type of intake best suited to the particular location must be made. Descriptions of the various forms of intake and the features of each type are given in Chapter 4 above.

5.1.5 Dimensions of Main Structure

The size of the structure is defined by the combinations of maximum demand and minimum supply water level at each season of the year. The area of the entrance to the structure through which the supply is passed is sized to give an acceptable value of maximum velocity - say 1-2 m/s in a gated opening with open channel flow; maximum velocities for lined power tunnels and gates in these tunnels would be higher and restricted by head loss, lining material, roughness, etc.. Where the openings are not gated, where it is intended to install screens, where it is desirable to avoid attracting fish or where turbulence is to be avoided (e.g. pump intakes) appropriate maximum velocities will be adopted.

The level of the sill at the entrance to a free surface conduit can be set lower than that of the entry sill (see 5.1.6 below) but consideration should be given to lowering the bed of the forebay so that heavier particles carried over the entry sill will be dropped before the water enters the main intake structure. In general the sill of the openings will define the level of the floor of the structure except where pumps are to be installed; in the latter case the floor inside the structure must be lowered to ensure that sufficient submergence for the pumps is provided under all operating conditions. Entrances to submerged power tunnels may have to be sufficiently deep to avoid vortex formation.

The sizing of the intake to a tunnel or pipe system will be restricted by acceptable velocities in the culverts. These velocities are associated with screen area or velocity through gate openings.

Detailed hydraulic calculations will be required to confirm the performance of any tentative intake design, and to ensure that it has an adequate capacity.

5.1.6 Entry Sill

An entry sill is usually provided upstream of the intake structure in order to prevent as much as practicable of the transported sediment in the river from reaching the intake. The level of the entry sill is critical. It should be set at such a height above the bed of the river that the bed load and the higher concentration of the suspended sediment are excluded at high river flows (see Figs. 2.1 and 2.2 in Chapter 2): but at the same time the crest of the sill must be low enough to enable the water demand to discharge over the sill at low river levels. In the latter circumstances the sill should be of sufficient length to keep the velocity to the required value.

If these requirements conflict then a decision must be made:-

(i) Accept a lower intake supply at low river levels,

or (ii) accept a higher sediment intake at high river flow,

or (iii) consider river training works to lower the river bed of the channel local to the intake,

or (iv) construct cross-river works to raise the river water level at low flows,

or (v) a combination of the above choices.

5.1.7 Arrangement and Orientation

The entry sill should be orientated parallel to the flow of the river to discourage deposition of bed load against the face of the sill. Where a weir or barrage has been constructed across the river to raise water levels it is often necessary to install gates in the weir immediately downstream of the intake to act as scour sluices and physically prevent build up of sediment in front of the intake (see Fig. 4.4). In such a case river training works, in the form of divide walls or groynes, across the front of the intake may be constructed to constrict the channel to the scour sluices and thus generate velocities locally exceeding those of the river prior to construction of the weir.

Although the entry weir is set parallel to the direction of flow of the river the axis of the forebay and the main intake structure should be set at a smaller angle to the river (as described in Chapter 3 above).

Where the intake takes the form of a tower in a reservoir the entry sill usually will be that of the intake gates. The lowest gate should be set clear of the anticipated deposition of sediment in the bottom of the reservoir, but below the normal operating drawdown level. If the operating range of the reservoir is considerable it is normal practice in water suppy systems (but not hydro-electric) to introduce alternative intake gates at vertical intervals of 10 to 20 metres providing adequate submergence so that water clear of both sediment and floating debris can be selected.

Where appreciable quantities of sediment are drawn over the entry sill but retained by a downstream sill, provision can be made for installation of scour sluices and a discharge channel from the forebay to control build up of sediment.

5.2 Geometric Recommendations

The boundaries of flow formed by the walls and floor of the approach to the main intake structure should be aligned with the intention of preventing turbulence and also of preventing separation of the streamlines from the boundary surfaces and inducing reverse eddies. The use of curved walls and nosings to piers will reduce turbulent effects. The prevention of separation of streamlines from boundaries is more of a problem. A rule of thumb used for canals is that the radius of curvature of the axis of flow should be at least two and a half times the width of the water surface of the channel. In practice such a large radius can seldom be provided at the approach to an intake and the wall at the inside of the bend will be curved to suit the space available. A nosing of radius 0.2 times the channel width will avoid the worst effects of flow separation.

To ensure even distribution of water across the face of the intake structure, floor baffles, often set at an angle to the flow, or vertical columns may be necessary. Even distribution of the water is of greatest significance where the intake contains pumps, or where the intake is set directly across a supply canal much narrower than the structure. Re-entrants and areas where slow moving water will cause the deposition of sediment should be avoided.

Some geometric recommendations are illustrated in Chapters 4 and 6.

For a large intake structure there is no substitute for hydraulic model tests to ensure satisfactory hydraulic performance, and the study of flow patterns which influence particle movements should be emphasised.

5.3 Floating Debris

Attention must be given to prevention of floating material from reaching the intake, and to disposal or dispersion of debris trapped against screens or baffles at the intake. Build up of debris, once started, proceeds at an accelerating rate and can throttle the water supply unless timely steps are taken to obviate the threat.

In reservoirs where water level ranges are large but velocities are low, a floating boom is a possible remedy. Such devices are often impracticable in rivers liable to high flows as they can be swept away during floods. They are also vulnerable to damage from logs or ice runs.

In rivers it is more common to attempt to deflect the debris at the entry sill to the intake. Where there is sufficient depth of water a deep skimmer wall mounted on splitter piers provides a submerged orifice for entry of water at all flows and forms a wall of sufficient height to prevent overtopping at high flows. In other rivers it is often necessary to resort to screens, usually of 50 - 150 mm bar spacing, which may be mounted on the entry sill if the river velocity is sufficient to assist in self cleaning. If the debris has to be cleared manually then the screens are usually installed at the openings to the main intake structure where access is easier and safer. Skimmer walls used with scour sluices as indicated in Fig. 5.1 have been used successfully. The sluice will have a gradient of about 1 in 3 with a gate on its downstream face to prevent the guides and spindle from choking. A radial gate for flushing should have its trunnion below the centre of curvature of the gate face to ensure a clean and rapid break from the silt packed against it. The trash sill if deeper could incorporate a crest gate for small flushes.

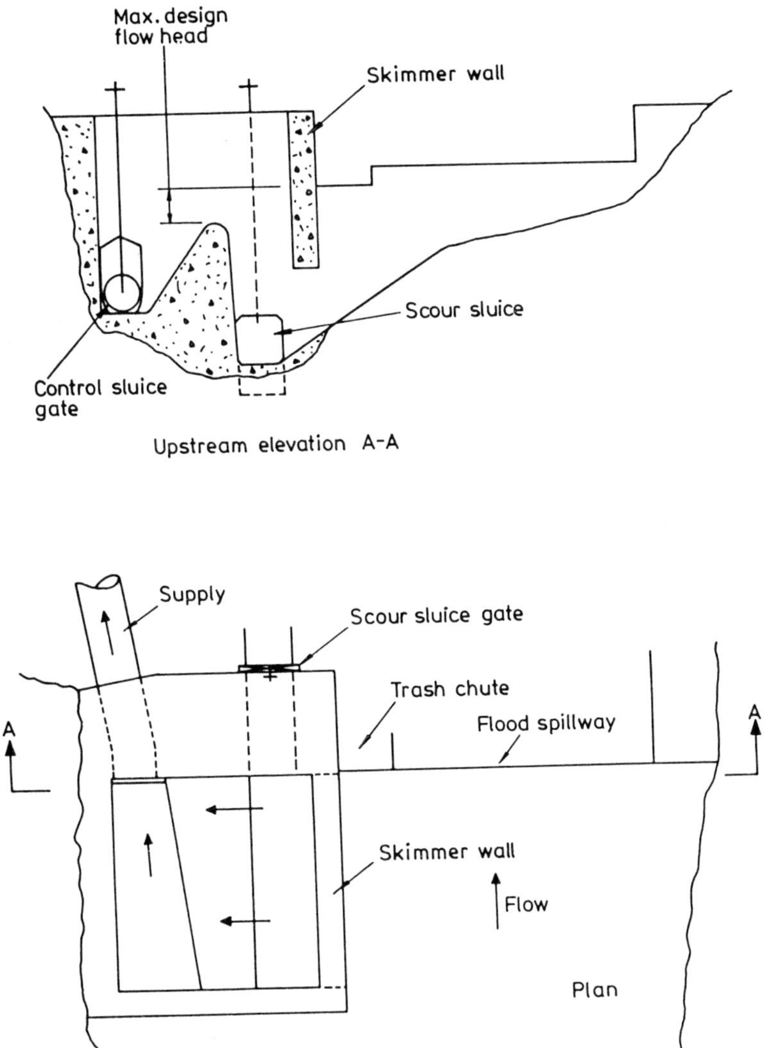

Fig. 5.1 Intake layout using baffle

The amount of debris liable to become trapped against the screens must be assessed. Regular manual raking - restricted to shallow intakes, say 5 m deep from deck level - may be sufficient, but large quantities of matted vegetation may require the use of powered mechanical rakes which may be arranged to operate automatically, by a timed or level sensing device. If virtually continuous accumulation of debris is anticipated then consideration should be given to drum screens, if the range of water levels is not too wide, or to band screens where the range is greater. Both these devices can be made self cleansing but require manual removal of the accumulating material. Disposal of trash should take into consideration other users downstream.

In cold climates ice presents special problems. Frazil ice may accumulate on screens and completely block them. Also, the build up of thick sheet ice can exert considerable pressures on the vertical faces of structures, and the usual remedy is to construct the walls with sloping faces so that the ice at the edge of the river is forced upwards. At time of thaw the presence of heavy ice flows in the river can pose a threat of considerable physical damage by impact; in such cases the geometric arrangement of the works must take account of the probable directions of movement of large pieces of ice in order to reduce the risk of direct collision.

5.4 Fish

In a number of countries both commercial and sporting interests are sensitive to the possible destruction of fish as a result of them being drawn into water intake works. A number of methods have been tried to prevent entry of fish and these include:

(i) use of fine mesh screens, in addition to the coarse screens which prevent entry of debris and protect the fine screens. These are very liable to blockage by the smaller debris and leaves so a cleaning gantry may be required. The velocity through the screens must be kept low so that fish are not trapped against the mesh. A maximum velocity of about 0.5 m/s is usually adopted. The mesh, where young fish have to be preserved, can be as fine as 4 mm spacing.

Where the fish screens are remote from the river channel it is important to avoid fish becoming trapped at the screens by ensuring that a route exists for them to return to the main river channel.

(ii) artificial creation of turbulence downstream of the intake to attract fish there rather than to the intake itself. An extension of this system is the use of vertical louvres leading past the intake to a fish bypass at the downstream end. For this to be effective the velocity of approach to the louvres needs to be approximately in the range 0.7 to 1.2 m/s.

(iii) use of electrodes suspended in the water above the entry sill and a ground conductor on the sill to form an electric screen. If the velocities are high in the region of the intake the stunned fish may still be drawn towards the intake.

The velocities and mesh sizes given above are only approximations and it is strongly advised that designers consult the fisheries authorities who will be associated with the intake structure and associated waterways.

Chapter 6

SEDIMENT EXCLUDING DEVICES

6. SEDIMENT EXCLUDING DEVICES

6.1 Curved Channel Sediment Excluder

This device uses the established practice of relying on the favourable effects of the curvature of a channel to reduce the amount of sediment entering the headworks of canal systems. The secondary currents in a curved channel which cause a large proportion of bed load to move towards the convex bank or wall of the channel are described in Chapter 3.

This classification is intended to include those sediment excluders for canal intakes where a separate curved channel is constructed from the river or main water course to the canal headworks which conveys the canal flow plus a sluicing flow. So it has the limitation of requiring sufficient quantites of water to allow some to be 'wasted' for sluicing purposes. This description is not intended to cover those cases in which the curvature of the whole river or main water course is enhanced by training works of one kind or another.

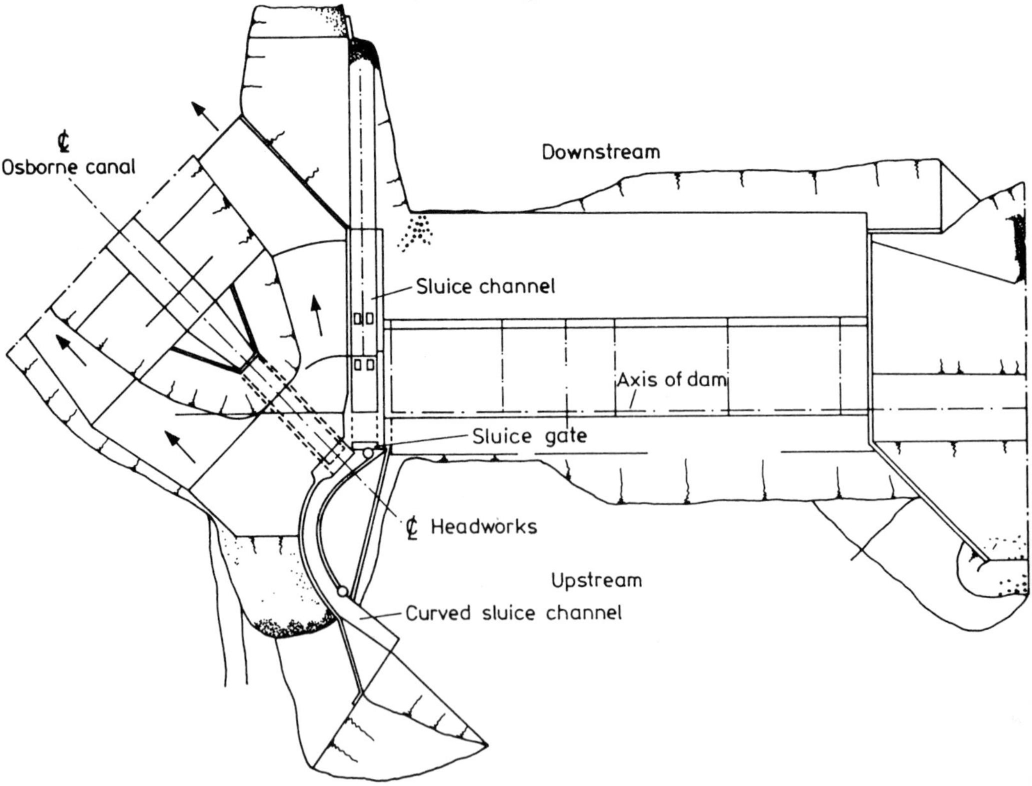

Fig. 6.1.1 Separate curved sluice channel for sediment exclusion (Osborne Canal)

There are two main types of curved channel sediment excluder: Type 1 in which the flow from the curved channel into the canal is through a relatively short, gate-controlled outlet, as in Fig. 6.1.1, and Type 2 in which the flow from the curved channel into the canal is over a relatively long skimming weir, as in Fig. 6.1.2.

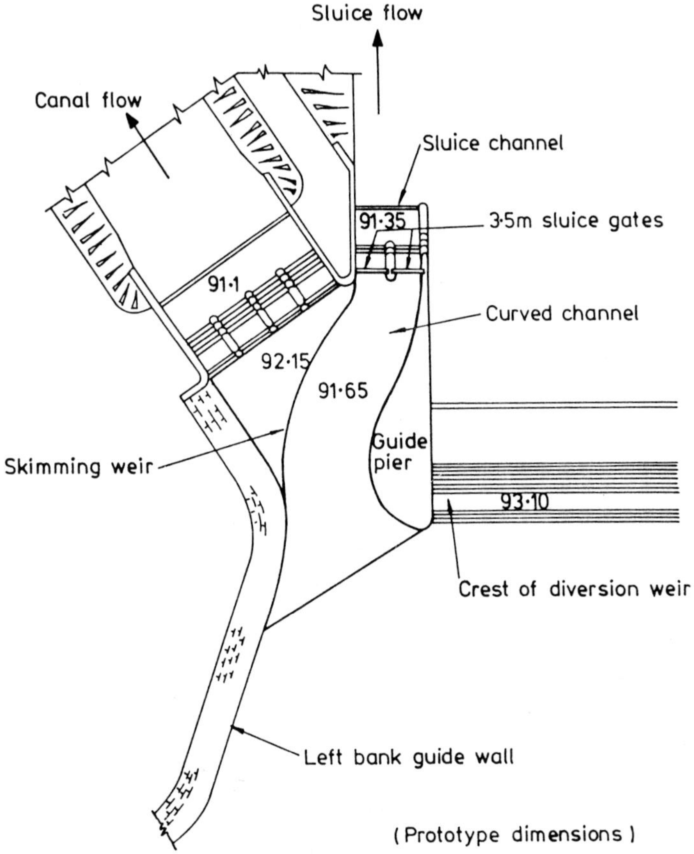

Fig. 6.1.2 20 m^3/s Headworks

The principle of the curved channel excluder is that sediment close to the bed moves to the inside or convex wall of the channel, the outlet to the canal being on the outer or concave wall of the channel. The amount of sediment entering the canal is thereby reduced. Some simple points must be taken into account: (a) the approach to the curved channel must be such that the desired helicoidal flow pattern does develop in the channel, (b) the canal entry arrangements do not cause sediment to be thrown up into suspension, and (c) velocities in the curved channel maintain the forward movement of sediment towards the sluice gates.

Since the flow in the curved channel decreases as it approaches the sluice gates, it is necessary to reduce the width of the channel in that direction to maintain the velocity so that sediment is neither deposited on nor entrained from the bed. The converging channel is shown in Fig. 6.1.2. A converging curved channel has some useful characteristics: (a) provided the curvature is not so severe that separation occurs, the flow is essentially irrotational, i.e. velocity is inversely proportional to the radius of curvature and (b) the line of maximum velocity shifts towards the convex bank where it may help in conveying concentrations of sediment towards the sluice gates.

Factors affecting design are summarised below:

(a) The intake location should be in accordance with Chapter 3 of this Guide. Obviously the intake should be located favourably with respect to river curvature and it is essential that the river or main channel water enters the curved channel in a streamlined and controlled way. This division of flow is as important as that between the curved channel and the canal and it is essential to ensure that a disproportionately high sediment concentration is not attracted to the intake works.

The total flow entering the intake structure will be of the order of 1.3 to 1.5 times the specified canal flow, Q_c. The width of the channel at the entrance to the intake structure should relate to the width of a channel in the main water course which would convey Q_c. The Regime method, which relates to channels with mobile boundaries, predicts a required width (in metres) of $5\sqrt{Q_c}$ (Q_c in m^3/s). It is suggested that from this width the curved channel is narrowed in a streamlined way until it reaches the dimensions required for the sluice channel.

(b) The water level in the curved channel will be determined by other civil works requirements. If the natural river levels are not suitable, it may be necessary to construct a structure across the river to achieve the necessary water levels in the curved channel.

(c) The bed level of the curved channel will probably be determined by the river bed level at the sluice gates; or possibly by the level at which it is considered that sediment will be swept away to avoid blockage of the outlet of the sluice channel.

(d) The sluice flow is usually fixed as a proportion of the canal flow, depending on sediment size and concentration. Values of around 30% and 50% are typical.

(e) Some guidance about plan geometry can be obtained from Table 1 (from Ref. 6.1.1) which gives the leading characteristics of five existing curved channel sediment excluders. It is suggested that the ratio of centre-line radius to the width of the curved channel at its mid-length should be between 3:1 and 5:1.

(f) It has been suggested that to achieve reasonable performance the Froude number in the curved channel should be in the range 0.5 to 0.8 at the upstream entrance to the canal head regulator.

(g) The conditions downstream of the sluice channel in the main water course must be such that blocking of the channel is avoided at all times.

The properties of the typical sediment will be an important consideration. At this point some consideration must be given to criteria determining the velocity of the flow in the curved channel. Obviously if a flow equal to something like 30% to 50% of the canal flow is to continue through to the sluice gates, then quite high velocities are to be expected in the curved channel. Possible criteria to consider are:

(a) Since it is intended that the curved channel does not accumulate sediment and since its floor is a plane surface, it could be supposed that the curved channel velocity would be satisfactory if the shear stress imposed on the sediment particles by the flow exceeded the critical shear stress, for example as determined by Sheild's criteria (Appendix 2, section A 2.3). An example of such procedure is given later. However, calculations on some existing curved channels indicate that the imposed shear stress is several times the critical shear stress, providing a reasonable safety factor. The ratio of imposed shear stress to critical shear stress may still remain a useful index of performance. Some recent experimental work suggests that a value of τ_* near the entrance to the curved channel of about 0.1 may be appropriate.

(b) The reality of the situation is that a sufficient velocity is required to ensure that a fairly high concentration of sediment is carried along in the flow. For example, if the concentration of sediment in the flow at the entrance to the headworks is 500 ppm, the concentration adjacent to the sluice gates could be 1000 ppm. The sediment carrying capacity of the flow can be checked by sediment transport formulae, or alternatively as in the example by a semi-empirical relationship which determines sediment transport as a function of major parameters of the flow and sediment particles.

The velocity corresponding to critical shear stress (τ_*) on the sediment particles can be determined as follows:-

(a) The average velocity, V_r, in the vertical at any radius r is assumed to follow the relationship V_r proportional to 1/r. This will result in a logarithmic distribution of the average velocities in the verticals in a section across the curved channel. If Q is the discharge at a particular section in the curved channel, then

$$\frac{Q}{d} = V_i r_i \ln (r_o/r_i) \qquad 6.1.1$$

where d is flow depth, V_i is the average velocity in a vertical close to the inner wall of curvature r_i and r_o is the radius of curvature of the outer, or concave, wall. From equation 6.1.1 the value of V_i can be determined, and from the proportionality relationship the average velocity in any vertical can then be determined.

(b) It is now possible to check the average velocity required to maintain critical shear stress on the sediment particles using the Shields' relationship (Appendix 2, section A 2.3) and a logarithmic mean velocity equation for flow in the channel. The most practicable method of using the Shields' criterion is probably that given in the ASCE Sedimentation Manual (Ref. 2.2), the relevant graphical relationship being reproduced in Fig. A2.2 (of Appendix 2). Firstly, it is necessary to calculate a dimensionless grain parameter I, where I is defined below

$$I = \frac{D}{\nu}\left(\frac{0.1\ gD\ (\gamma_s - \gamma)}{\gamma}\right)^{\frac{1}{2}} \qquad 6.1.2$$

Then from Fig. A2.2 the dimensionless critical shear stress τ_*, can be read off where:

$$\tau_* = \frac{\tau_o}{(\gamma_s - \gamma)D} = \frac{v_*^2}{gD(\frac{\gamma_s - \gamma}{\gamma})} \qquad 6.1.3$$

(since $v_* = \sqrt{\tau_o g/\gamma}$)

Knowing the numerical value of the left-hand side of equation 6.1.3, the value of the shear velocity v_* can be calculated.

Finally the mean velocity corresponding to critical shear stress can be calculated from an equation such as:

$$V = 2.5\ v_*\ \ln\ (\frac{12.3\ R}{\Delta}) \qquad 6.1.4$$

where Δ, the effective roughness of the bed is usually taken as the 65% size of the bed material but is subject to a correction factor if the bed is not completely rough.

(c) The velocity determined using equation 6.1.4 can be compared with the actual velocity in the curved channel.

(d) Alernatively the value of τ_* calculated from equation 6.1.3 can be compared with possible necessary values as suggested previously. Assuming that a preliminary design has been completed, assisted by the procedures outlined above, it would be advisable to carry out a model study of the excluder. This is because the regime of flows in the river, and the variability of the hydrograph, are very significant factors in the performance of a sediment excluder. In some cases it is possible that for extended periods, flow is run into the canal headworks even though little water is available for sluicing. Simulation of the river flows and the operation of the sluice gates will give some insight into sediment exclusion performance under these conditions. Performance under flood conditions can also be investigated.

The method used in the example in Appendix 3, for determining sediment concentration is the graphical relationship developed by Colby (Ref. 6.1.2) as reproduced in Fig. A2.4.

Table 1 - Leading characteristics of some curved channel sediment excluders

Headworks	Canal flow m^3/s	Sluice flow m^3/s	Sluice width m	Centreline radius m	depth m	Average Angle to intake centreline
Courtland	11.3	5.66	6.0	-		
Superior	2.26	1.13	7.0	-		
Republic	3.39	1.69		-		
Bartley	1.70	1.06	3.6	11.0	1.73	60°
Woodston	1.19	0.99	1.5	7.6	2.59	90°

6.2 Vortex Tube Sediment Extractor

A vortex tube sediment extractor is a device for the continuous removal of sediment moving near the bed of a channel. It consists of a horizontal tube or duct installed normal to and below the bed of the channel, which extracts a small proportion of the flow near the bed, where generally there is a higher concentration of bed material load. A horizontal axis vortex is generated in the tube and the flow and sediment are conveyed laterally to a settling basin or discharge channel. The vortex tube can be located in the approach to the canal headworks or sufficiently far downstream of the headworks to ensure that the equilibrium of the sediment distribution in the canal is established.

The comments in the section apply to vortex tubes located in canals downstream of the headworks, where the flow is subjected to more controlled conditions and vortex tube performance is more predictable than upstream of the headworks. A possible disadvantage is that the intake has to be sized for abstracted flow plus waste flow.

The flow enters the vortex tube tangentially and generates a forced vortex along the axis of the tube. The flow through the tube is controlled by a gate at the downstream end where it discharges into a disposal channel. A typical example of a vortex tube installation is shown in Fig. 6.2.1.

Vortex tubes are most appropriate where substantial bed load is to be excluded and have limitations with respect to sediment distribution and suspended sediment. However, appreciable trapping efficiencies have been recorded at sites where suspended local is predominant.

6.2.1 Design of the Vortex Tube

A considerable amount of research on the theory and performance of vortex tubes has been carried out by the Hydraulics Research Station, Wallingford. This was based initially on the analysis of previous research and on hydraulic models of vortex tubes. This work was later confirmed and refined by field measurements on prototype vortex tubes constructed recently in Indonesia and Nepal. The measured trapping efficiencies at these extractors were about 76% and 50% respectively, with median bed material sizes of 0.38 mm and 0.2 mm.

For the hydraulic design of the vortex tube it is recommended that Refs. 6.2.1 and 6.2.2 should be used to calculate the optimum tube dimensions; and the method given in Ref. 6.2.3 should be used to predict the trapping efficiency.

The design objective is to remove enough sediment from the canal, such that, the remainder does not exceed the sediment transport capacity in the downstream canals. The design method involves the assembly of canal and sediment data and the selection of trial tube dimensions and limits on acceptable extraction ratios (vortex tube discharge to approach canal discharge). This is followed by trial hydraulic calculations from which acceptable combinations of tube dimensions, operating head and extraction ratio can be selected. The trapping efficiencies are then computed to determine the optimum design, capable of removing the required amount of sediment from the canal to satisfy the downstream canal sediment transport capacities.

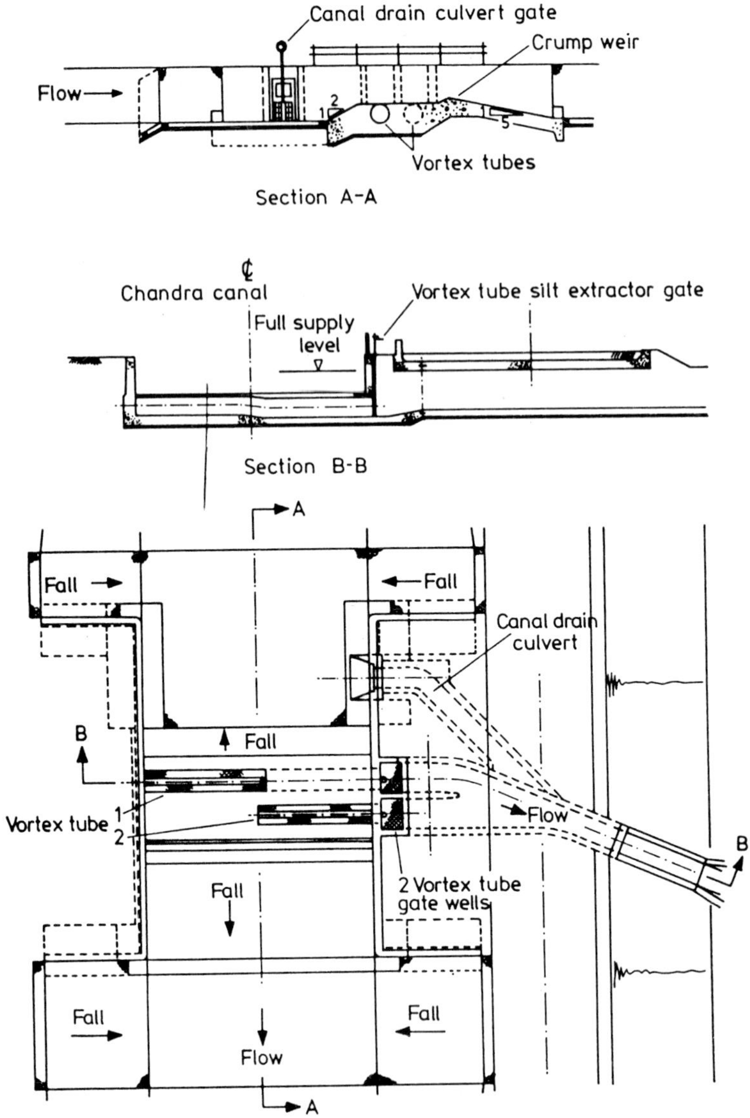

Fig. 6.2.1 Vortex tubes

6.2.2 Level Setting for Vortex Tube

The level at which the tube is set in relation to the bed of the canal, effects the Froude number in the canal at the vortex tube. Generally for finer sediment low Froude numbers give higher trapping efficiencies, while with very coarse sediment the Froude number can be increased. One advantage of increasing the Froude number, by raising the vortex tube, is that, although it does not affect the overall available head, it reduces the possibility of backing up in the vortex tube disposal channel, at times of high downstream river levels.

6.2.3 Location of Vortex Tube

Ideally, the location of the vortex tube should be as near the headworks as possible to optimise the removal of the sediment. To ensure reasonable equilibrium of sediment distribution in the canal profile, it is recommended in Ref. 6.2.1 that the distance (in metres) downstream of the headwork should not be less than: 10/Fr x Vn/Vs50.

where:- Fr - Froude number in approach channel

Vn - Normal channel velocity at vortex tube (m/s)

Vs50 - Settling velocity of median size sediment (m/s)

In many cases where there is insufficient head for the disposal of trapped sediment it may be necessary to move the vortex tube further downstream.

6.2.4 Disposal of Trapped Sediment

In common with most sediment removal devices, the main problem encountered with vortex tube sediment extractors is the difficulty in the disposal of the trapped sediment. For the disposal of the sediment back to the river or escape channel, by gravity, it is necessary to provide for the following head losses:-

- difference in head between canal full supply level and minimum canal supply level.
- operating head for vortex tube.
- head losses through the downstream of vortex tube control gate.
- head loss in disposal channel.

These losses should not exceed the difference between the full supply level in the canal at the vortex tube and the maximum river level at the disposal channel outfall, when the canal is operating. The disposal channel should be as short as possible and the outfall located at a point in the river, where there is a high concentration of flow; for example on the outside of a bend. Backing up of the flow within the disposal channel, should be avoided.

Because of high sediment concentrations, the ideal solution for the design of the disposal channel is to provide for supercritical flow. Failing this, the channel design should be checked for sediment transport capacity (see Appendix 2).

In many cases it is inevitable that compromises will have to be made. Where possible, provision should be made for high extraction ratios, for flushing out the sediment during periods of river flows in excess of canal demand. Alternatively a flushing sluice can be located in the canal upstream of the vortex tube.

The alternative to gravity disposal is to provide a settling basin and mechanical or pumped removal of sediment. The smaller flows from the vortex tube extractor, compared with a canal settling basin, tend to favour the use of compact settling tanks and pumped disposal of the sediment.

Where there is a shortage of water supplies for continuous ejection, an alternative solution is to provide settling tanks with intermittent pumped disposal of the sediment and re-use of the vortex tube flow. For this purpose it is necessary to locate the vortex tubes upstream of a canal fall. This arrangement serves the same purpose as a dredged settling basin but eliminates the more skilled requirements for the operation and maintenance of the dredgers, and controls the size of material trapped.

6.3 Side-Sluice Sediment Excluders

Applications:

Side-sluice sediment excluders are used to abstract quite large flows from rivers carrying appreciable quantities of sand and gravel. They require a small head (fall provided by a weir across the river) to operate and are easily cleared if blocked during a flood.

In common with all sediment excluders, side-sluice sediment excluders do not affect suspended sediment. In what follows, sediment should be taken to mean sediment travelling along the bed.

PRINCIPLE OF OPERATION

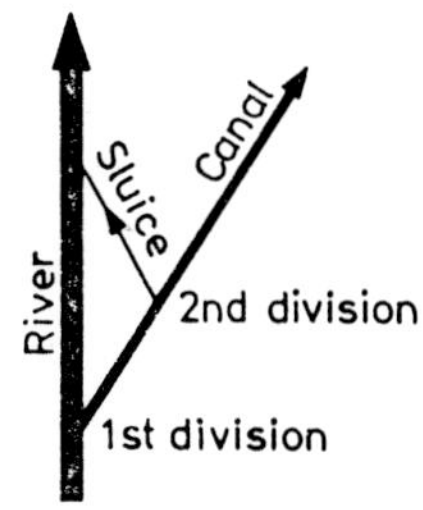

A side-sluice sediment excluder works by dividing the already diverted flow into two parts, one on either side of a vertical dividing wall. One part is returned to the river through a sluice channel and the other is passed into the canal and used. Fig. 6.3.1 illustrates the principle.

Fig. 6.3.1 Flow diagram for right bank intake

The twin objects of the design procedure are (a) that the sluice water should carry a higher concentration of sediment, and the canal water a lower concentration, than the water initially diverted, and (b) that the water initially diverted should have a lower concentration of sediment than the river itself. There is no point in achieving one of these objectives if it is at the expense of the other, because the overall reduction in concentration is the product of the reductions at the two divisions of flow.

DESIGN OF INITIAL DIVERSION

The design of the initial diversion is important when only part, Q_d, of the river discharge, Q_r, is being diverted, because their relative sediment concentrations will be affected.

The discharge Q_d is the sum of the canal and sluice discharges Q_c and Q_s and can for design purposes be taken as:

$$Q_d = 1.5Q_c \qquad 6.3.1$$

where Q_c is the maximum canal discharge.

The discharge Q_d, when in the river and approaching the headworks, occupies a certain width, b_d, out of the total width of flow b_r, and it is sufficiently accurate to assume that

$$\frac{b_d}{b_r} = \frac{Q_d}{Q_r} \qquad 6.3.2$$

The total width of flow at any discharge depends on factors such as how long that discharge has been flowing or whether it is rising or falling, but it may roughly be taken as

$$b_r = 5\sqrt{Q_r} \quad (b_r \text{ in metres, } Q_r \text{ in m}^3/\text{s}) \qquad 6.3.3$$

although any strong evidence in a particular case may indicate the use of a factor other than 5.

Combining equations 6.3.2 and 6.3.3 gives

$$b_d = 5\,\frac{Q_d}{\sqrt{Q_r}} \quad (b_d \text{ in metres, } Q_d \text{ and } Q_r \text{ in m}^3/\text{s}) \qquad 6.3.4$$

For a chosen value of Q_r, the dividing streamline upstream of a given diversion structure can be sketched in. If the dividing pier is further out into the river than b_d, the dividing streamline will curve away from the bank and vice versa.

The basic hydraulic principle at work is that bed currents cross a dividing streamline from the outside to the inside, thereby directing the bed load either towards or away from the intake, depending on the direction of curvature of the dividing streamline. At the same time surface currents cross the streamline from the inside to the outside. The two currents are equal, which is what makes the line an average streamline in plan.

In Fig. 6.3.2 more surface water is diverted than bottom water, with the result that sediment concentration in the diverted water is reduced. In Fig. 6.3.3 the exact opposite is true.

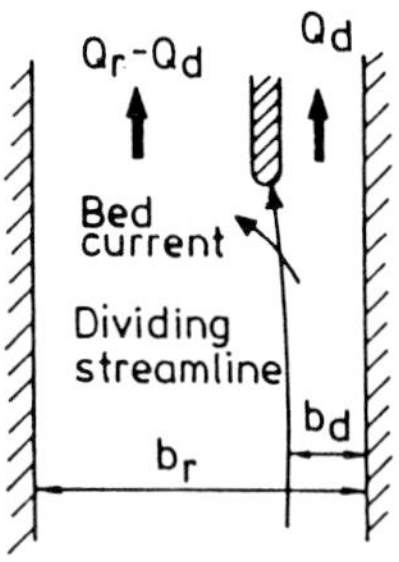

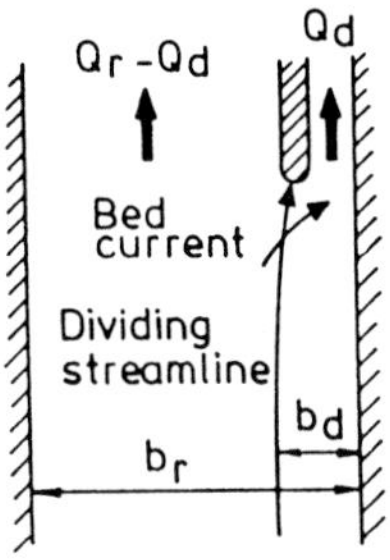

FIG. 6·2·4

Fig. 6.3.2 **Fig. 6.3.3**

Plan view of approach to right-bank intake

Now equation 6.3.4 shows that as Q_r increases so b_d decreases. A layout that falls in Fig. 6.3.2 when Q_r is only just greater than Q_d will do so at any higher value of Q_r also. It is therefore sufficient to set the centreline of the pier out from the bank a distance, b_e, equal to b_d when $Q_d = Q_r$, giving

$$b_e = 5 \sqrt{Q_d}$$

$$b_e = 6 \sqrt{Q_c} \qquad 6.3.5$$

The above treatment is based on an important proviso: for a right-bank intake the width of flow in the river is always measured from the right bank. In other words, it is assumed that whatever the discharge, the flow 'hugs' the bank in which the intake is located - this is generally the case only where the intake is located on the outside bank of a bend in the river (3.1).

DESIGN OF SLUICING ARRANGEMENT

The same principles apply to the design of the second division of the flow into the sluice water and canal water, Fig. 6.3.4.

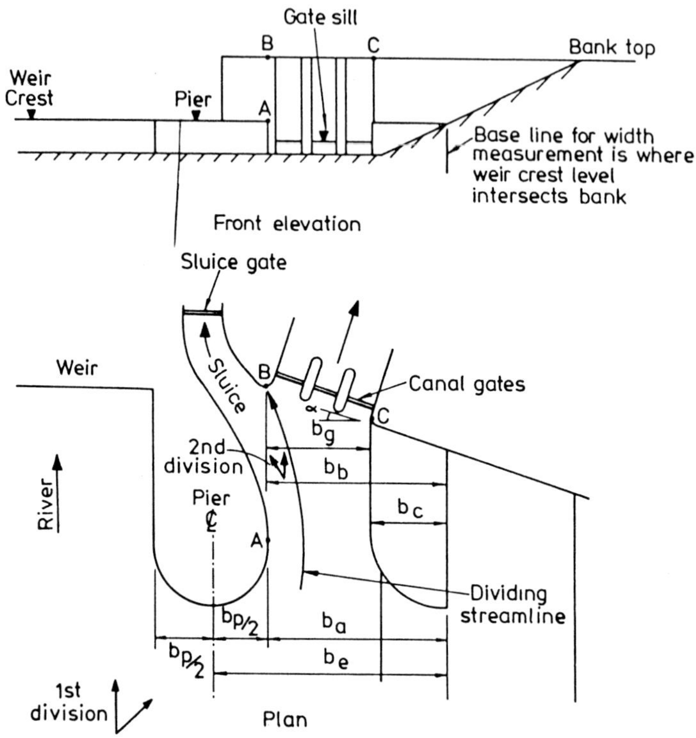

Fig. 6.3.4 Right bank intake definition sketch

If, looking downstream, the point B is further from the bank than point A, then there will be no value of the sluice discharge, Q_s, for which the dividing line will not curve the right way. It is sufficient to make the line AB parallel to the bank, giving

$$b_b = b_a = b_e - \frac{b_p}{2} \tag{6.3.6}$$

The canal gates have to resist the pressure of the river in flood and are a costly item. Their width can be reduced by moving B closer to the bank, which means increasing the width of the pier (equation 6.3.6), by keeping the angle α down to around 20 or 30°, and by moving C out from the bank while shaping the bank to remove the resulting sharp corner at C.

A well proportioned structure arises if

$$b_p = \frac{b_e}{2} \text{ to } \frac{2b_e}{3} \qquad 6.3.7$$

and $$b_c = \frac{b_p}{2} \qquad 6.3.8$$

resulting in a frontal width for the gates equal to half or a third of that resulting from a thin pier and no bank outstand. A complete design along these lines is shown in Fig. 6.3.5.

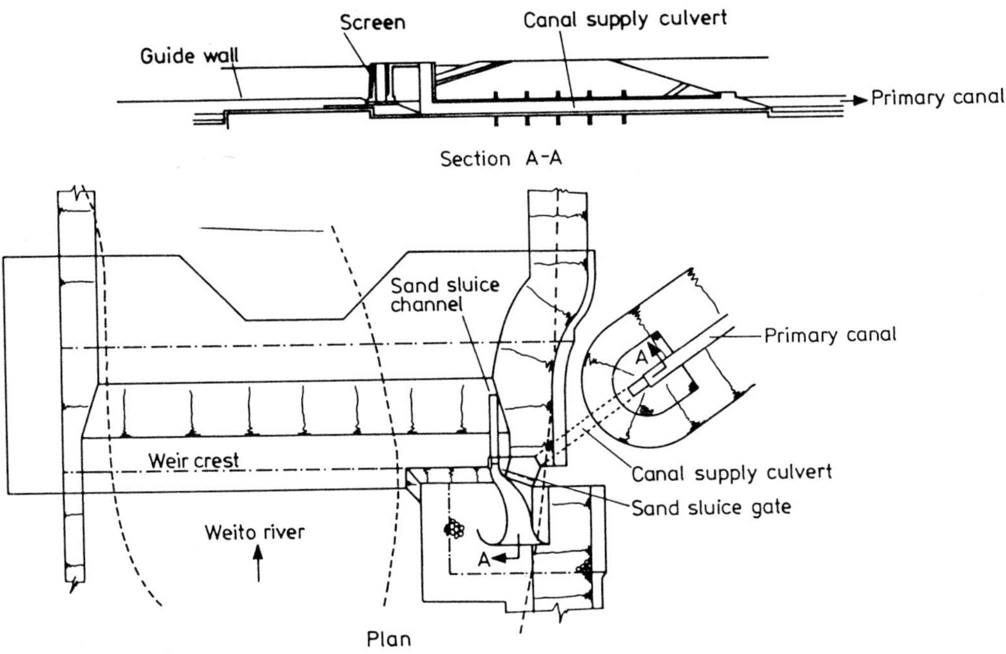

Fig. 6.3.5 Weito irrigation project - proposed diversion weir and headworks

Summary of Widths (units in metres and m^3/s)

Q_c = canal capacity

$$b_e = 6 \sqrt{Q_c} \qquad 6.3.5$$

$$b_p = 3 \sqrt{Q_c} \text{ to } 4 \sqrt{Q_c} \qquad 6.3.9$$

$$b_c = \frac{b_p}{2} \qquad 6.3.8$$

$$b_g = b_e - b_p \qquad 6.3.10$$

HEIGHT OF PIER

If the pier is built up to bank top level, floating trash in times of flood will get trapped by it in front of the gates. If the pier is built up to weir crest level only, it will be completely effective when no water is passing over the weir, and will continue to be effective even when moderately submerged. At higher flows the pier, which is now only an extension upstream of the weir crest, becomes largely invisible to the water, which sees only the solid outstand from the bank formed by the gates. Since this outstand at high flows intercepts a much higher discharge than is actually taken into the gates, very favourable conditions of the Fig. 6.3.3 type cause vigorous scouring in front of the gates and prevent any accumulation of sediment. Note that these are the requirements of a plain intake, which is what the present intake becomes at high river flows.

FLOOR LEVELS

The floor level of the sluice channel and its narrowest width should be such that a discharge Q_s equal to $Q_c/2$ or more can pass down it without backing the water up to such a level that it starts to flow over the weir. This calculation must take account of the tailwater level to be expected at the sluice channel outfall. If the tailwater level resulting from Q_s being discharged into the river is so close to weir crest level that the sluice channel must be made wide and deep in order to pass the flow, shear stress or sediment transport calculations are necessary in order to check whether sediment will settle on the floor of the sluice channel.

The whole of the floor between the pier and the bank and in front of the gates may be set at the same level as the floor of the sluice channel. Some designers feel the need to have the floor in front of the gates at a higher level, with the resulting step, shown in Fig. 6.3.5, called a skimming weir. It is important to realise that a step in the floor can only help in as much as it produces the correct curvature in the dividing streamline, on which alone the direction of the bed currents depends, and the effect of a step in this respect is not at all obvious. If the direction of flow near the floor is up and over a step, sediment soon forms a ramp against the step which the following sediment is able to ascend. This happens, for instance, on a short stretch of step near point B, if the curvature of the sluice channel is inverted too soon after passing point B.

FLOATING TRASH

An intake designed to exclude bed sediment in the way described automatically attracts any floating debris that may be in the river. The two are inseparable parts of the same process. This can be a problem in rivers with a lot of trash, and easy access from above should be provided to the areas in front of the intake.

The hydraulic details of the design of such an excluder must be sorted out using physical or mathematical models.

6.4 Tunnel Excluders

The principle of tunnel excluders and ejectors is similar to that of frontal intake structures in that the flow approaching a barrage is split horizontally and the clearer water flows through the upper chamber into the canal, while the sediment laden water flows into the lower channel and is passed back into the stream. A typical tunnel excluder is shown in Fig. 6.4.1. This type of device has proven very efficient for large and small diversions as a sediment excluder - used upstream of the diversion point - and as an ejector - used downstream of the diversion point.

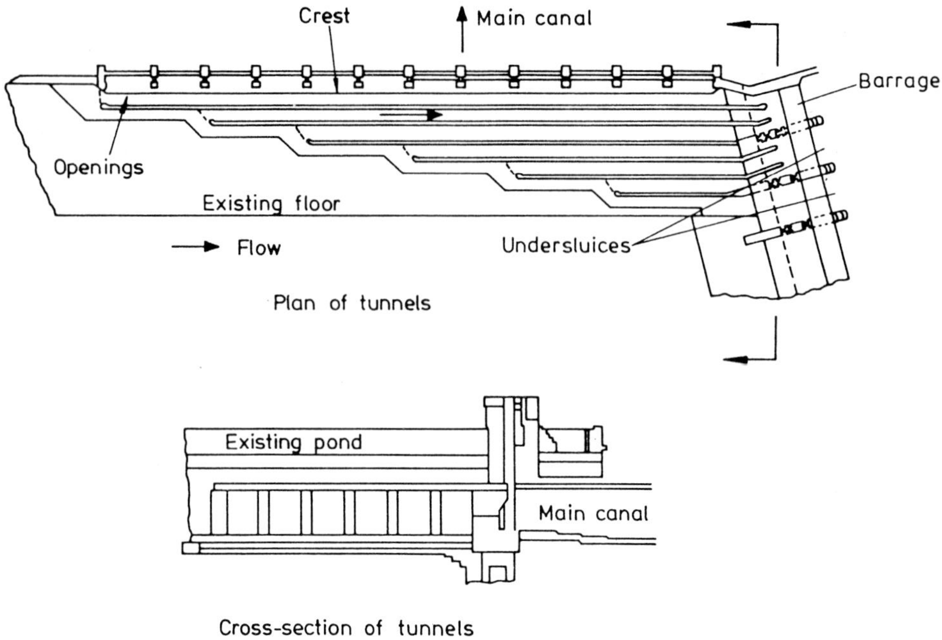

Fig. 6.4.1 Tunnel type sediment excluder

Literature reviewed in Ref. 3.1 give a number of examples of tunnel excluders and some design guidelines:

(i) Entry to the tunnels should be bellmouthed to avoid turbulence where the flow is split.

(ii) Approach flow direction should be straight and the site should be carefully chosen to ensure that the tunnels can cope with incoming discharges.

(iii) The approach, sizing and entry conditions should be physically modelled to ensure that the tunnels will operate efficiently.

(iv) Downstream conditions must be carefully considered to ensure that the downstream ends of the diverting tunnels are kept clear of deposits at all times.

6.5 Approach Flow Control

Although approach flow control is clearly not a sediment excluding device, it is well established that control of approach flow can achieve sediment exclusion from intakes. The aim is to provide the favourable effects of channel curvature at the approach to the intake to prevent bed load sediment from entering the intake.

If natural channel curvature is insufficient or ill-defined at the intake structure it may be induced by the construction of guide walls, curved vanes, etc. upstream of the intake. Examples of walls and vanes to control flow are numerous and a number are included in Ref. 3.1.

When a barrage is constructed across a waterway and intakes to canals are to be positioned on both banks, the effects of unfavourable curvature are almost bound to influence at least one of the water systems to be supplied. To improve the situation a central island and guide banks may be used to provide favourable curvature to both banks - Fig. 6.5.1.

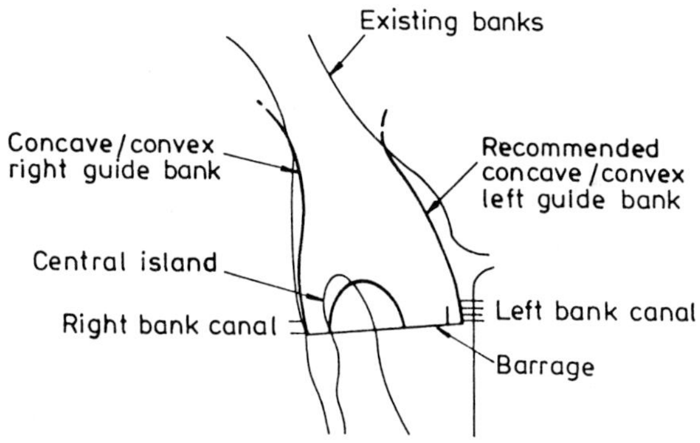

Fig. 6.5.1 Example of alignment for guide banks using central island to induce channel curvature

Literature on approach flow control suggests that intakes should always be constructed where suitably curved approach flow is available, and if this is impossible, curvature should be brought about artificially. However, it has been shown that where tunnel excluders (section 6.4) are used, maximum efficiency is obtained when the approach flow is straight. So control of the approach flow must be considered fully, in light of the other aspects of the design. When guide walls, vanes, etc. are to be used their location, extent, curvature, etc. should be investigated using a model study where the whole range of operating conditions can be studied. For instance the ratio of flow approaching the barrage and the intake depending on gate openings can greatly influence the quantity of sediment conveyed in either direction.

6.6 Sediment Excluding Intakes

The sediment excluding features of a number of intake structure types and the features of some sediment excluding devices have been described in chapter 4, and earlier in this chapter. Many intakes incorporate combinations of these methods to exclude and control sediment. A few intake designs developed with sediment exclusion and control as a main criteria are described here.

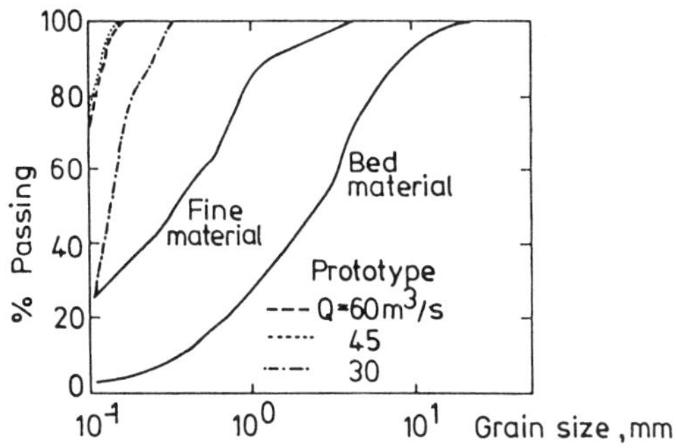

Fig. 6.6.1 Grain size distribution curves

Existing problems due to sediment entry at the intake on the Kander River, Switzerland instigated model tests to redesign the intake (Ref. 6.6.1). A number of alternatives were considered and tested. A maximum river discharge of 185 m^3/s was assumed and the sediment grain size distribution is shown in Fig.6.6.1. The resulting intake design, shown in Fig. 6.6.2, combined the use of approach flow control and a tunnel excluder just upstream of the canal proper to control sediment.

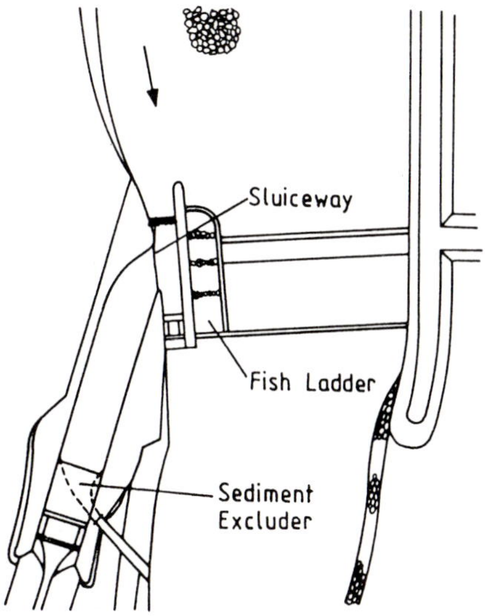

Fig. 6.6.2 Final layout of intake

The total flow diverted is Q = 22 m^3/s, with Q_c = 16 m^3/s being taken in the canal. For these conditions, sediment will enter the intake for flows between Q_o = 24 m^3/s (the incipient motion discharge) and Q = 45 m^3/s. This corresponds to entry of sediment over 96 days of the year. Initially, the sediment that does enter deposits in the mouth of the sluiceway. Subsequent sediment is moved to the right side at the entrance of the channel leading to the excluder. However, the second curve re-distributes the sediment reasonably evenly over the whole width so that flow and sediment conditions are uniform at the excluder. This double use of curvature within the intake (see Fig. 6.6.2) was found to be the most effective means of attaining uniform flow conditions. Considering scale effects, the material deposited in the canal for different flow rates shows that only silts and fine sand fractions will escape the excluder in the prototype. A trash rack is located at the entrance of the channel leading to the excluder. The gate at the end of the sluiceway will be used to enable sluicing of material from in front of the intake and in the sluiceway, and to control water levels within the intake so that material deposited in the secondary intake channel can be flushed through the excluder.

The performance of the excluder can be improved by employing a regulation scheme where the excluder is shut for flows less than Q = 20 m^3/s i.e. when no sediment transport will occur. It was considered that management of the intake would be simplified if the excluder was left functioning for all flows greater than Q = 20 m^3/s. For all low flows, the water level will be kept at the crest level of the weir.

For flows greater than Q = 45 m^3/s, the spiral flow created by the groyne in the river bed and the curvature of the right bank keeps the transported bed load away from the intake.

Two mandatory requirements of the project were that a residual flow of 0.7 m^3/s be left in the river for low flow periods, and that a fish ladder of acceptable design be incorporated into the weir or intake. The fish ladder can be seen in Fig. 6.6.2. The flow down this structure simultaneously satisfies the residual flow requirement.

Another example of new intake design is that described in Ref. 6.6.2. The aim of this work was to develop sediment control facilities for diversion headworks in mountain streams. The design which was developed is shown in Fig. 6.6.3. Instead of horizontal separation of the flow, a slot in the bottom of the intake is used to separate sediment and water. Underneath the slot the sediment is collected in a sediment trap and flushed downstream where it is returned into the river. The hydraulic principle is similar to that of a vortex tube sediment extractor used sometimes in irrigation canals. However, the application in connection with a diversion and the integration in an intake structure was new and created additional challenge.

This design for sediment control at diversions proved to be a well functioning alternative to existing conventional devices. It can be applied whenever the plan area is limited and sufficient head is available. Considering these presuppositions it is obvious that the method is mainly restricted to diversions in mountain rivers with high slopes and steep banks.

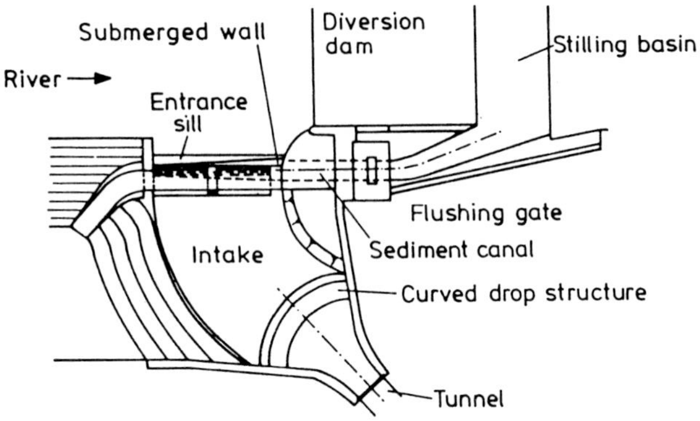

Previous design

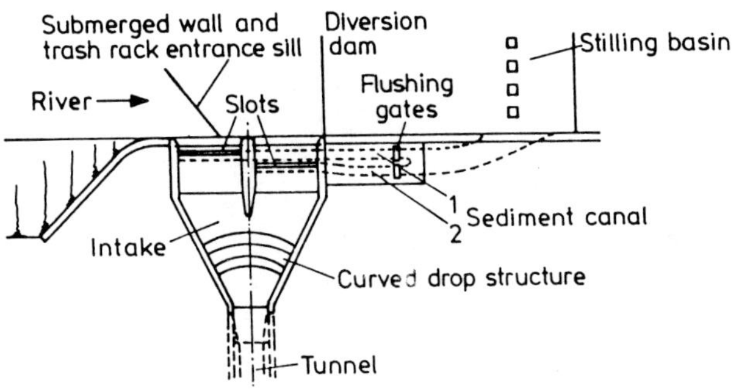

Final design

Fig. 6.6.3 Subersach diversion

For proper functioning several construction rules and operational instructions have to be observed carefully. With respect to the sensitivity of the new intake type against unfavourable flow pattern in the vicinity of the intake and against wrong operation of the flushing gates, the carrying-out of hydaulic model tests, by which the design was developed, is highly recommended.

The Brander No 1 intake for the Awe Project of the North of Scotland Hydro-Electric Board was designed to control sediment ingress. The intake incorporates a sediment settling channel and a water operated sector gate for scouring. Fig. 6.6.4 shows scouring in operation.

Fig. 6.6.4 Cruachan aqueducts - Brander (1) intake - scouring

<u>References</u>

6.1.1 ICID Ninth Congress on Irrigation and Drainage, Moscow 1975.

6.1.2 Colby, B.R. "Discharge of sands and mean velocity relationships in sand bed streams", US Geological Survey Professional Paper 426-A, Washington 1964.

6.1.3 Russein, A.S.A. Experimental study of a curved channel sediment excluder - experimental work carried out at Chilworth Hydraulics Laboratory, University of Southampton.

6.1.4 Developments in Hydraulic Engineering, Volume I, Novak, P. Editor (Chapter 5, K.V.H. Smith, Developments in the Design of Irrigation Structures), Applied Science Publishers, London, 1983.

6.2.1 Sanmuganathan, K. "Design of vortex tube silt extractors", Report OD/6, Hydraulics Research Station, March 1976.

6.2.2 White, S.M. "Design manual for vortex tube silt extractors", Report OD37, Hydraulics Research Station, December 1981.

6.2.3 Atkinson, E. "Field verification of a performance prediction method for sediment extractors". Report OD 90, Hydraulics Research, August, 1987.

6.3.1 Blanchet, C. "Importance of location of water intake on a river for prevention of entry of sediment" (in French). Paper R22 in Vol. 2 (Q30) Trans. I.C.I.D. 9th Congress, Moscow, 1975.

6.6.1 Whittaker, J.G. "A solution for sediment control at intakes". Proc Channels and Channel Control Structures 1st International Conference, Southampton, England, April 1984.

6.6.2 Scheuerlein, H.J. "A new design of sediment control facilities for diversion headworks in mountain streams". Proc Channels and Channel Control Structures 1st International Conference, Southampton, England, April 1984.

Chapter 7

SETTLING BASINS

7. SETTLING BASINS

Where a river carries a substantial quantity of sand and finer particles in suspension, this material cannot practically be excluded at the intake. A settling basin, in which the velocity of the offtaking flow is reduced to enable the sand and heavier silt load to settle out under gravity, is used in such circumstances to reduce the sediment load in the downstream system.

The river intake installations where settling basins are commonly used are as follows:-

(i) In irrigation schemes, to reduce the sediment load to a level which can be mainly transported by the distribution system through to the fields, thus mitigating the problems of sediment deposition in the canals. (The sediment transport capacity of irrigation canals is often restricted due to the flat water surface slopes needed to maintain command.)

(ii) In power schemes to reduce the sediment load to acceptable levels from the viewpoint of (a) sediment transport capacity of the supply system to the power station and, (b) damage to machinery.

(iii) In water supply schemes to reduce where necessary the sediment load in the supply by gravity sedimentation as a preliminary to subsequent chemically-assisted sedimentation and/or filtration.

The application of gravity sedimentation in water supply schemes is limited because impurities such as algae, vegetative debris, fine silt and colloids do not settle out adequately under gravity without the use of flocculants. Note also the separate requirement for bankside storage in river abstraction schemes (typically 3-7 days' capacity) where the river may be subject to industrial pollution - such reservoirs cannot strictly be termed settling basins although as a secondary function they may settle out considerable quantities of sediment.

In this chapter the basic theory and approach to settling basin design is summarised, and the practical aspects involved in their layout are reviewed. The scope is limited to gravity (as opposed to chemically-assisted) sedimentation - i.e. to settling out discrete particles which retain their individual settling characteristics without interference or flocculant effects. Discussion is limited to horizontal flow basins as typified by the enlarged canal cross-section in irrigation systems, or the rectangular tank common in water supply schemes (Fig. 7.1). The basic settling theory is also applicable to the more sophisticated vertical-flow and radial-flow tanks found in water supply installations (Ref. 7.1)

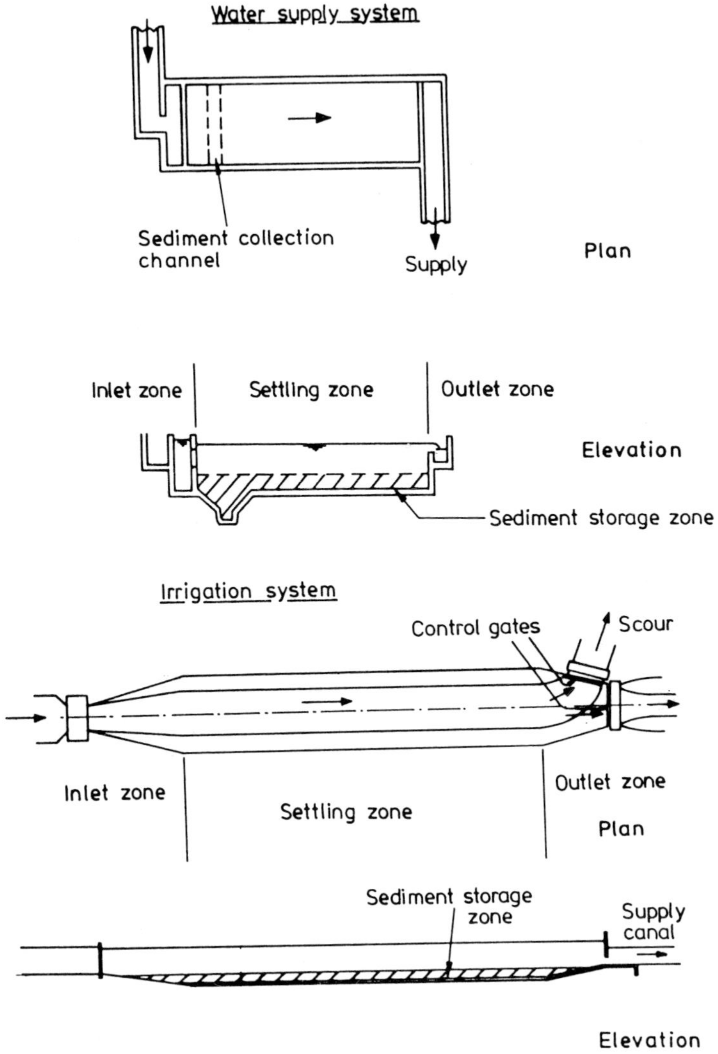

Fig. 7.1 General layout of settling basins

7.1 Fall Velocity of Discrete Particles

Fall velocity, w, in quiescent water characterizes the ability of different sized particles to settle out under gravity (and indeed is commonly used as standard measure of particle size for fine sediments). Fall velocity for discrete particles is dependent on particle size, specific gravity, particle shape, and the viscosity of water. Fig. 7.2 shows fall velocity in water, w, plotted against particle diameter, D, for reference quartz spheres. Various equations exist which give approximate solutions for fall velocity of single particles - the Rubey equation (Ref. A2.11) is commonly used for particles with the shape of natural sands. The ASCE Manual on Sedimentation Engineering (Ref. 2.2) can be consulted for further details. The significant effect of water temperature on fall velocity - particularly in relation to design for the tropical environment - should be noted.

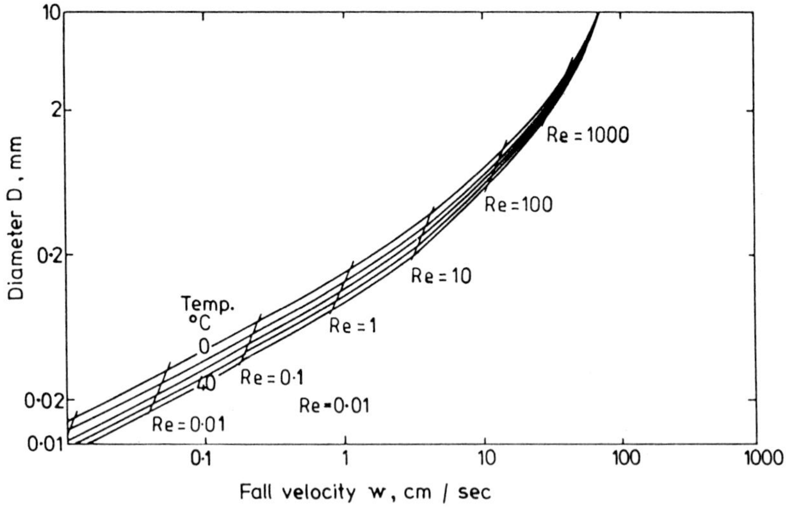

Fig. 7.2 Fall velocity of quartz spheres in water

Effect of Concentration on Fall Velocity

With increasing concentration of sediment, actual particle fall velocity will differ from that for discrete particles due to interference of other particles. Flocculation can occur in high concentrations of silt, clay and organic particles when the particles coalesce to fall in a group at a higher velocity (Refs. 2.2, 7.2). Removal of such fine particles is however generally outside the scope of gravity sedimentation - Miller (Ref. 7.3) provides a good review of the settling of suspensions from the viewpoint of secondary treatment processes in the water supply industry. Hindered settling occurs when discrete particles settle in close proximity to one another, and their velocity fields interfere. Interference effects become significant at suspended sediment concentrations in excess of 2,000 mg/l (Ref. 7.2), when the reduction in fall velocity for a coarse silt suspension might be about 10% (Ref. 2.2). In general, the effects of hindered setting are not significant in terms of the ranges of sediment concentrations and the degree of accuracy considered here for settling basin design.

Determination of Settling Characteristics of Suspended Sediment

Design fall velocity for discrete particles may be measured directly by timing fall through a known depth. It is usual, however, to use published data (such as Fig. 7.2 or Ref. 2.2) to estimate fall velocity when the basin is principally concerned with settling out sand and coarse silt sizes. In water treatment processes, where finer particles need to be removed, it is usual to carry out a settling column analysis to determine the settling characteristic curve for the suspension (Ref. 7.2).

7.2 The Ideal Basin

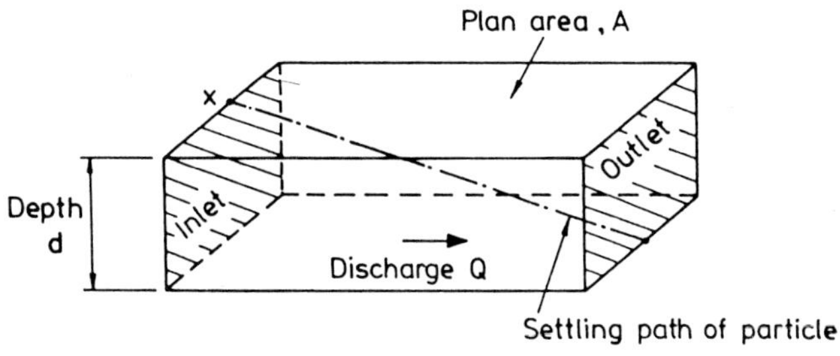

Retention time, t_R = Ad/Q

Settling time, t_S = d/w

Fig. 7.3 Ideal settling basin

The ideal horizontal settling basin, Fig. 7.3, demonstrates the basic theory of sedimentation developed by Hazen (Ref. 7.4). The following assumptions are made:- uniform distribution of flow and suspended solids at entry to settling zone (plug flow); quiescent flow (i.e. no turbulence); solids entering deposition zone are not resuspended. Consider a sediment particle entering the basin at point X :

Settling time, $t_S = d/w$ 7.1

Retention time, t_R = basin volume/discharge = dA/Q 7.2

where yo = basin flow depth; A = mean plan area of basin; Q = discharge. For quiescent settling, all particles of settling velocity w are removed when retention time equals settling time:

$$\text{i.e.}\quad dA/Q = d/w, \quad \text{or} \quad Q/A = w \qquad 7.3$$

A similar formula can be obtained for a vertical flow tank. In general for both ideal and real basins, the ratio $\frac{wA}{Q}$ can be regarded as a dimensionless indicator of the physical ability of a basin of plan area A to remove particles of fall velocity w at supply discharge Q.

It follows, in the ideal case, that for discrete particles:

(a) removal is independent of basin depth and flow-through velocity,

(b) for a given discharge and suspended sediment load, removal is a function of basin surface area (note in this context the use of compact double-storey tanks in water treatment).

The ratio Q/A is termed the "surface loading" or "surface overflow rate".

7.3 Settling Capability of Real Basins

In practice, real settling basins act less efficiently than the ideal due principally to the effects of (a) turbulence in flow through the basin leading to retarded settlement, and (b) short-circuiting and currents within the basin. Sediment removal efficiency, η, for a given particle size is measured as C_r/C_o, where C_r = concentration of suspended sediment removed, and C_o = incoming concentration.

Hazen (Ref. 7.4) attempted to account for the effects of both turbulence and short-circuiting by a general classification of basin performance in his formula:

$$\eta = 1 - [1 + m (\frac{wA}{Q})]^{-\frac{1}{m}} \qquad 7.4$$

where m is a performance parameter varying from m = 0 for "best" basins to m = 1 for "very poor" basins. Hazen's equation is shown graphically in Fig. 7.4 and is still commonly quoted today. The disadvantage of Hazen's formula is that several different physical effects are combined into a single parameter, m. It is better for the designer to consider each effect separately as follows.

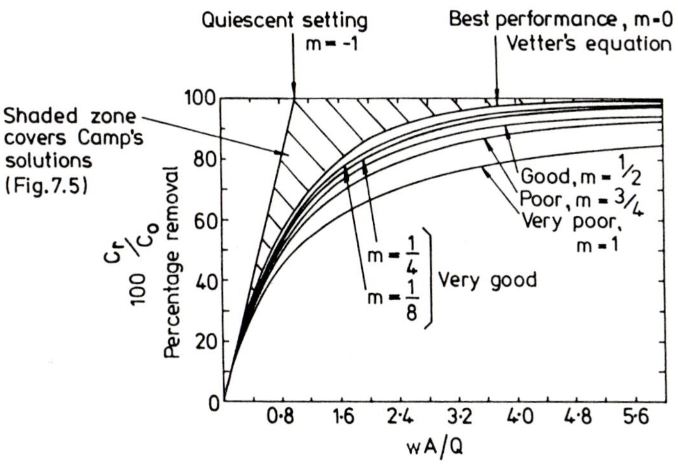

Fig. 7.4 Performance curves for settling basins of varying effectiveness

Effects of Turbulence

Camp (Ref. 7.5) based his classic approach to settling basin design on the work of Dobbins (Ref. 7.6). After making simplifying assumptions that (a) fluid velocity, and (b) the turbulent mixing coefficient are the same throughout the fluid, Camp derived a relation for:

$$\eta = f (\frac{wA}{Q}, \frac{w}{v_*}) \qquad 7.5$$

where v_* is the shear velocity and $\frac{w}{v_*}$ can be regarded as a dimensionless indicator of the effect of the fluid turbulence on a given particle size.

Camp's solution to equation 7.5 is shown graphically in Fig. 7.5 (the horizontal axis has been redrawn in terms of $\frac{w}{v_*}$).

$$\text{Shear velocity, } v_* = \sqrt{g\ RS} \qquad 7.6$$

where R = hydraulic mean depth, and S = hydraulic gradient which is calculated from a boundary resistance equation such as Manning's and essentially depends on flow-through velocity.

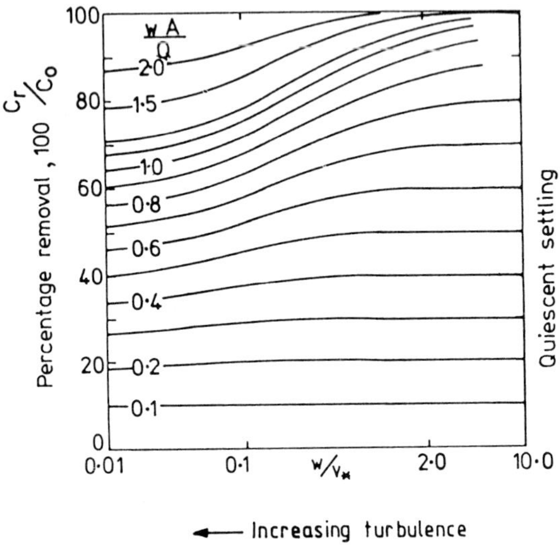

Fig. 7.5 Camp's solution for settling basin efficiency

The following table demonstrates how, for fixed flow depth and surface loading, flow-through velocity has to be reduced if the same scale of turbulence effects on settling is to be maintained when design particle size is reduced:

Design particle size, D (mm)	0.06	0.02
Mean flow-through velocity, V (m/s)	0.2	0.02
Hydraulic gradient, S	3.6×10^{-6}	3.6×10^{-8}
Shear velocity, v_* (m/s)	8.4×10^{-3}	8.4×10^{-4}
Fall velocity at 20° C, w (m/s)	3.5×10^{-3}	3.5×10^{-4}
w/v_*	0.42	0.42

Notes: d = 2.0 m and $V = d^{2/3}\ S^{\frac{1}{2}}/n_m$ where Manning n = 0.015.

The practical upper limit of flow-through velocity for basins designed to settle out sand particles is generally taken as 0.3 m/s.

Other recent approaches (Refs. 7.7, 7.8) have taken into account more accurately the effects of turbulence on settling. Camp's solution however remains widely accepted. In irrigation design practice, Vetters' equation (Ref. 7.9) is often used and is virtually identical to the equation proposed by the USBR (Ref. 7.10):

$$\eta = 1 - e^{-\frac{wA}{Q}} \qquad 7.7$$

This is simply the "best performance" solution of Hazen's equation, (i.e. curve for m = 0, Fig. 7.4). Note that Vetter's equation also corresponds to the turbulent side of Camp's solution (Fig. 7.5) and thus to implicit conditions of turbulence. Note also that Camp's solution could be plotted on Fig. 7.4 as a series of curves, similar to the Hazen performance curves. It can thus be appreciated that each of the common settling efficiency formulae (equations 7.4, 7.5 and 7.7) is valid when the implied physical conditions are appropriate.

Effect of bed scour

Once particles have settled out, they must not be scoured from the basin floor by excessive flow-through velocity. The shear stress on the floor must therefore be less than the critical shear stress required to initiate movement.

$$\tau_o = \gamma_w RS = s_w v_*^2 \qquad 7.8$$

where τ_o = bed shear stress, γ_w = specific weight of water, and s_w = specific gravity of water. The critical shear stress to initiate motion can be obtained for the D_{50} sediment deposit size from Shield's diagram (Fig. A2.2). Equating the two shear stresses enables the critical flow-through velocity to be obtained. Various equations giving particular solutions for critical flow-through velocity relating to the different bed conditions in Shield's diagram have been published (Refs. 7.5, 7.11).

Short circuiting and basin stability

In the ideal basin, flow is steady and uniform (plug flow), and all fluid particles are detained in the settling zone for the retention time, t_R. In practice, even with well designed basins, flow is non-uniform and some parts of the basin volume are ineffective. Lengths of stream paths of individual fluid particles vary - some reach the outlet in less than the theoretical retention time, while others take longer to do so. The flow-through curve for a tank (Fig. 7.6) provides a convenient indicator of hydraulic behaviour and efficiency. It illustrates the departure from ideal (plug) flow caused by short-circuiting of individual flow paths through the basin. The objective of good hydraulic design is to achieve conditions most closely relating to ideal flow.

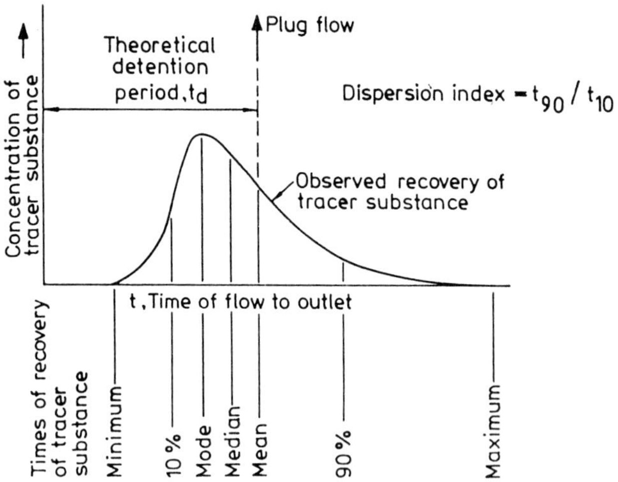

Fig. 7.6 Flow through curve - (Dead spaces and short-circuiting in a settling basin are reflected in the concentration and time of recovery of tracer substances).

Non-dimensional plots of the flow-through curve (Refs. 7.5, 7.12) enable comparison of the hydraulic behaviour of different basins. The degree and nature of short-circuiting determines the shape of the flow-through curve (Refs. 7.2, 7.5, 7.11). Also, if the flow-through curve for an existing or modelled basin does not reproduce itself reasonably well in repeated tests, then flow through the basin is unstable, and performance will be erratic.

Factors which cause short-circuiting of flow, with consequent reduction in hydraulic efficiency are:-

- currents set up by poor inlet and outlet conditions, and basin shape,
- dispersion in the horizontal plane due to turbulence,
- wind-induced surface currents,
- density currents induced by thermal effects.

The hydraulic design of the inlet and outlet layout and basin shape are the most significant from the design viewpoint - these are discussed in section 7.4. The other effects become increasingly important as flow-through momentum (i.e. velocity) is reduced - hence their greater relevance in the water treatment sector. Clements (Ref. 7.13) notes that the effects of wind can be reduced if it is possible to align the basin along the prevailing wind direction.

7.4 Hydraulic Factors Affecting Basin Layout

Settling Zone

Having determined the surface area of the basin, the designer must consider its plan shape. Camp (Ref. 7.14) demonstrated that the hydraulic behaviour (i.e. shape of flow-through curve) of long narrow tanks is superior to that of wide low velocity tanks, also that tanks with higher (but nevertheless very low) values of Froude number have better flow patterns and give less dispersion (Fig. 7.6).

In practice therefore, a minimum length to width ratio L/W of 2-3 is generally adopted from hydraulic considerations - note that in the water supply and power sector, the site layout frequently requires basin length to be minimized. In irrigation systems, it is generally more practicable to achieve a better L/W ratio (say 8-10) by local widening and deepening of the canal cross-section. Basin shape can be improved by subdivision with longitudinal divide walls, which may also be required due to operation considerations (section 7.5).

Inlet Zone

The need for a good inlet design cannot be overemphasised - poor inlet design is probably the factor most responsible for "poor" basin performance in Hazen's classification (Fig. 7.4). To achieve good hydraulic efficiency and effective use of the settling zone, the inlet strictly needs to distribute inflow and suspended sediment uniformly over the vertical cross-sectional area of the settling zone.

Clements (Ref. 7.12) has shown that horizontal velocity variations across the width of a rectangular tank affect the hydraulic efficiency considerably more than velocity variations in depth, provided always that bed scour is avoided. Principal attention therefore needs to be given to uniform inflow distribution in the horizontal plane. Methods which are commonly adopted to achieve good flow distribution are:

- submerged weir,
- gradual open channel expansion (Fig. 7.1), possibly using guide vanes,
- troughs with slots or orifices in walls or bottom, and
- baffle walls.

Orifices or baffled inlets are generally only used when extremely low flow-through velocities are needed for water treatment. As a general rule, the inlet layout should either follow an existing proven design, or be model tested.

Outlet Zone

The operating water level of the settling basin is generally controlled at the outlet, usually by a weir which may be designed to operate as submerged in order to conserve head. In irrigation and power systems, conventional undershot lift gates are also commonly used. If it is narrower than the basin, the outlet control requires an appropriate approach transition to avoid short circuiting and to maintain an even flow distribution. The outlet contraction may be more abrupt than the inlet expansion (Fig. 7.1).

7.5 Operation of Basin and Removal of Sediment

The method of removing sediment deposits from the basin must be considered early in the design procedure (section 7.7) since it may well be a critical factor governing the layout of the whole intake site. Removal of sediment may be either continuous (i.e. carried out during normal basin operation without interference to supply from the basin) or intermittent (in this context, carried out as a separate operation while supply from the basin is temporarily stopped). When considering removal of sediment, it is important to provide for removal from any part of the system where sediment is likely to deposit (e.g. dredger access to inlet transition) - conversely, it is equally important that adequate sediment carrying capacity is provided anywhere else in the system (e.g. basin approach channel, sluiceway channel) where sediment deposition is to be avoided.

In continuous systems, sediment is generally removed by hydraulic dredger (Fig. 7.12), or by mechancial scraper to a central sump where it is withdrawn either under basin hydrostatic head or by pumping (Fig. 7.13). Mechanical scrapers are generally only used in rectangular tanks for water supply systems. In intermittent systems, the basin is emptied or drawn down, and sediment is removed by hydraulic sluicing, by mechanical or manual excavation; or by pumping to a suitable disposal point. Environmental or river shoaling considerations may dictate the extent, method and frequency of disposal of sediment back into the parent river.

Wherever possible, gravity sluicing should be adopted since this is more effective and obviously cheaper than other means. Lack of adequate head may however prohibit it, or limit its effectiveness.

Provision of head for gravity sluicing can be an important secondary advantage of low-lift pumped abstraction from the river.

A sediment storage zone must be provided beneath the settling zone in any intermittent removal system. It is common for settling basins to be constructed in parallel so that any one basin can be taken out of service intermittently for removal of sediment, while continuous delivery is maintained by the others. The system of separate sedimentation channels shown in Fig. 7.7 is common in irrigation systems.

Gravity Sluicing

For effective gravity sluicing, the system is required (a) to erode progressively all deposits from the storage zone (Fig. 7.7), and (b) to convey this material at a high transport rate through the basin and associated sluiceway channel to the disposal point - generally the parent river. To achieve this, the fall through the removal system should be such that it operates at supercritical flow in the sluicing mode (Refs. 7.15 (R32, R35), 7.16). Further requirements are the provision of low level scouring sluices, and careful hydraulic design to ensure that no control to sluicing flow exists downstream of the storage zone. The mechanism of removal is shown in Fig. 7.7 - the design case being removal of the last remaining deposits.

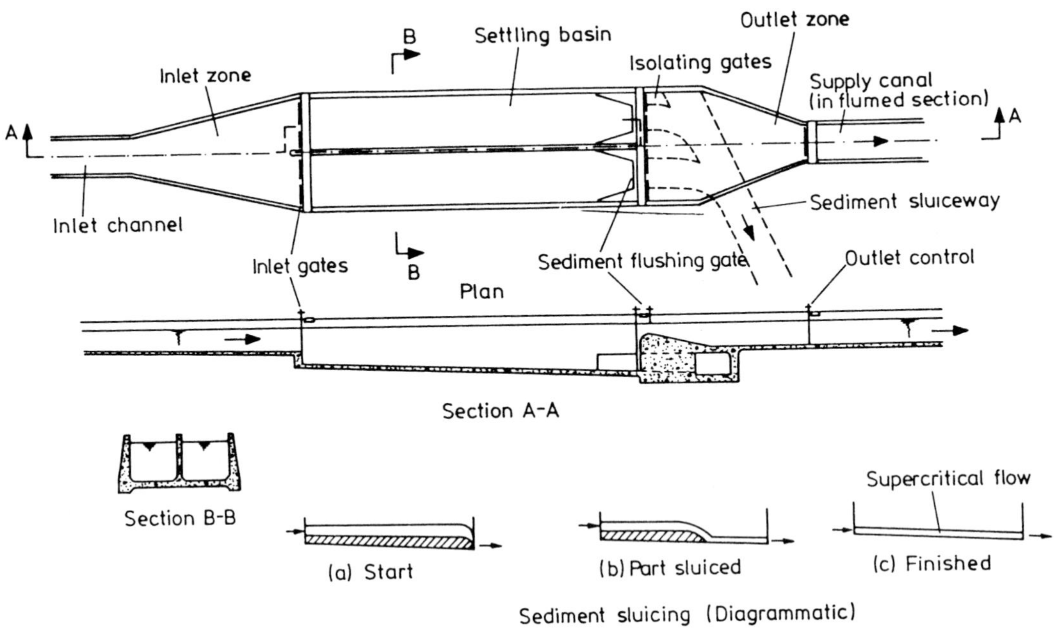

Fig. 7.7 Settling basin with parallel sedimentation channels

For removal of coarse silts and sand, it is suggested that sediment transport rates are estimated using Bagnold's total load formula (Ref. 7.17 and equation 7.9) since this is valid for high transport rates rather than the early stages of sediment movement.:

$$g_{si} = \gamma_w dSv \left(\frac{eb}{\tan\alpha} + \frac{0.01\ V}{w}\right) \qquad 7.9$$

where g_{si} = transport rate of solids by immersed weight per unit width; eb = bed load efficiency factor (Fig. 7.8); and tan α = solid friction coefficient (Fig. 7.9). Individual particle terminal velocity should be estimated for D_{50} sediment deposit size. Note however that Bagnold's theory is inapplicable for D_{50} less than 0.015 mm.

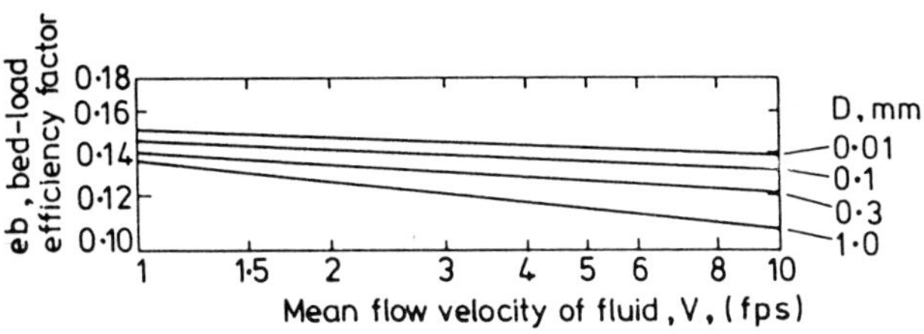

Fig. 7.8 Bagnold's bedload efficiency factor

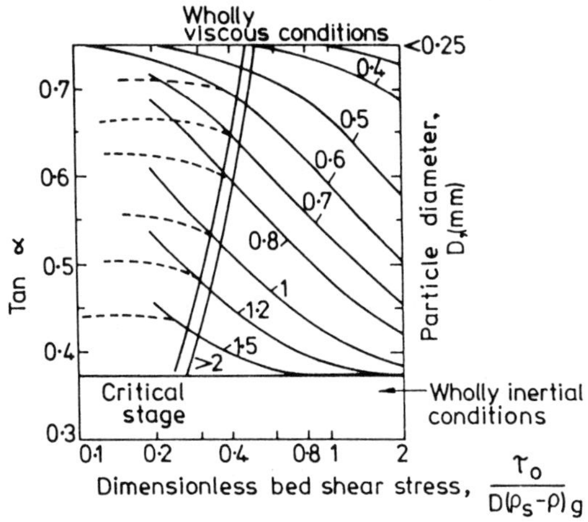

Fig. 7.9 Bagnold's solid friction coefficient

In many instances when designing a settling basin for low head river diversion works, there are generally conflicting design requirements such as:

- The head available for gravity sluicing of the settling basin.

- Lack of sediment storage capacity.

- Seasonal limitation of sluicing supplies.

- Difficulty in sluicing settling basin during long periods of high flood levels.

In many cases sluicing has to be confined to the lower range in flood levels, and sediment storage has to be sacrificed to produce sufficient sluicing power. The reduction in the storage capacity results in the more frequent shut down of the canal distribution system. Some of the difficulties can be overcome if the settling basin and flushing sluices are duplicated - as in Fig. 7.7. However, this adds considerably to the cost and requires additonal water supplies which may not always be available, and significantly it does not solve the problem of not being able to sluice the basins during long periods of high flood levels in the river.

An alternative means of overcoming these conflicting requirements is to provide for both settling-scouring operation and continuous ejection operation. During periods of low river flows and low sediment concentration, the normal operation of allowing the sediment to deposit in the settling basin, is followed by the scouring operation of the desanding sluices to remove the sediment. However, during periods of high sediment concentrations and high flood levels this method of operation may not be possible, and to remove the lower sediment laden water the desanding sluice gates are operated to allow for the continuous ejection of the sediment back to the river. When considering continuous sediment ejection, it is important that the ejection flow should pass through tunnels starting off parallel to and under the canal flow. An example of a settling basin and desanding sluices designed on this basis are shown in Fig. 7.10.

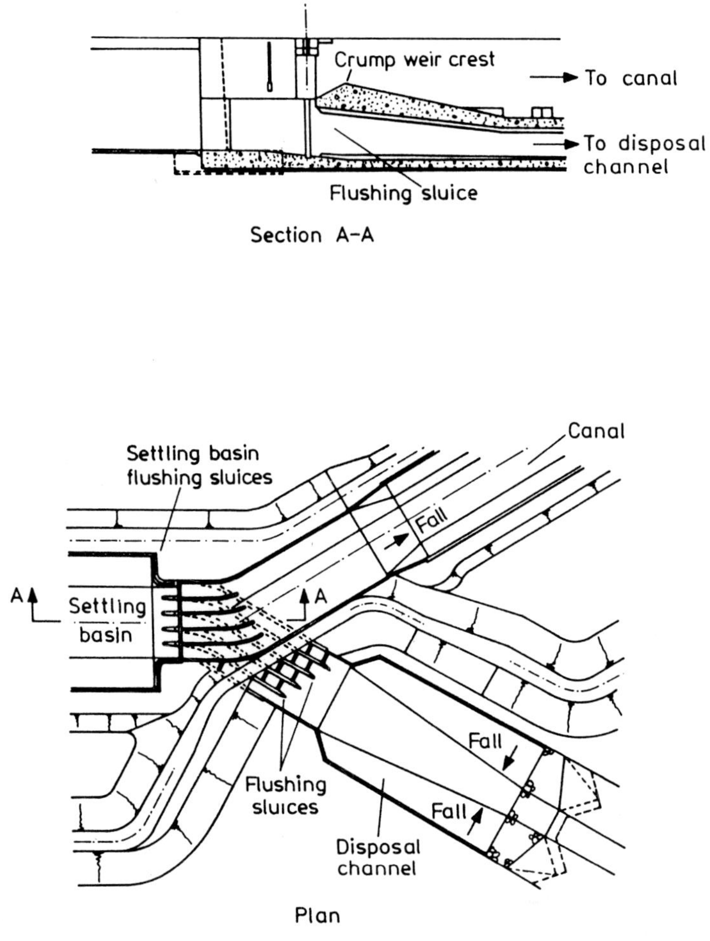

Fig. 7.10 Settling basin

7.6 Examples of Settling Basin Design

Relatively little prototype data has been published on the performance of settling basins outside the water supply field. Ref. 7.15 contains some general details, including recent data on the desilting works at Imperial Dam on the Colorado River (Fig. 7.11) - possibly the largest in existence.

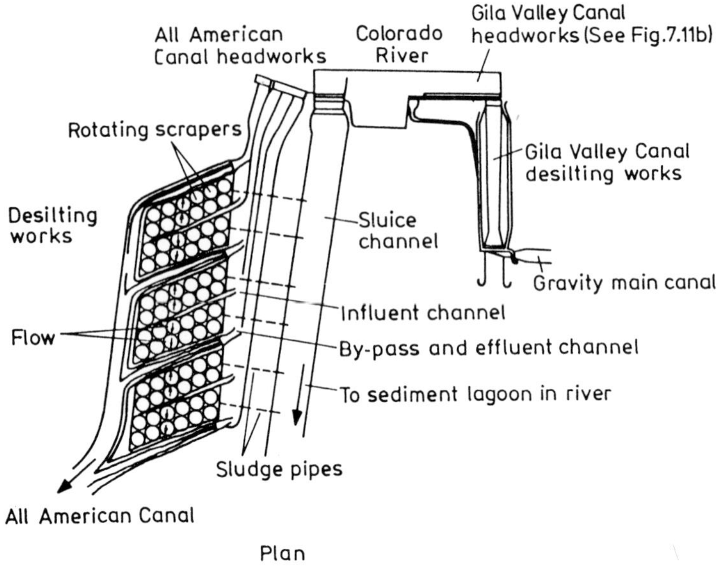

Fig. 7.11(a) Imperial dam desilting works

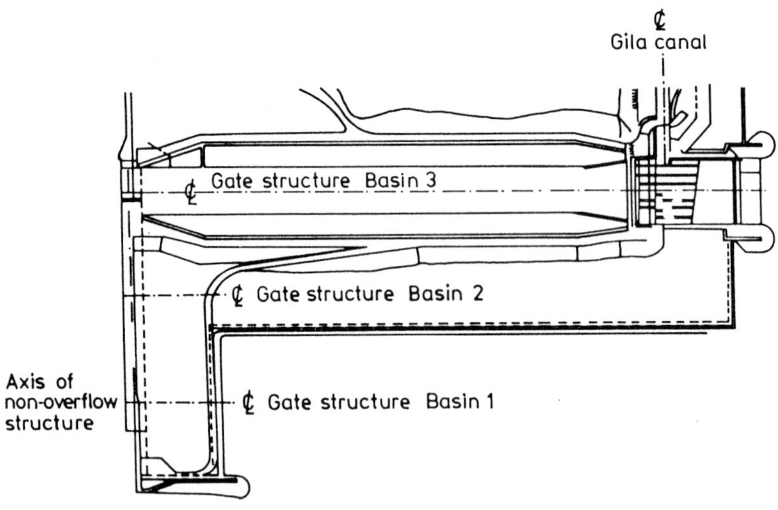

Fig. 7.11(b) Desilting works in Gila Valley Canal

The Imperial Dam works are designed for removal of suspended sediment particles larger than 0.05 mm from the supply to two large irrigation canals. Basin size was designed using Vetters' equation (equation 7.7). The right bank (All-American Canal, constructed pre-1940) desilting works comprise an unusual layout of three basins, each containing 24 no. 38 m diameter rotary scrapers. On each scraper, diagonal scraper blades move sediment continuously towards the central pedestal where the sediment is drawn off by hydrostatic head through a sludge pipe and discharged into the sluiceway channel. The surface area of each basin is 33,500 m^2, giving a surface loading of 3.4×10^{-3} m/s at the design discharge of 113 m^3/s. Design flow-through velocity is 0.08 m/s. The left bank (Gila Canal) desilting works, with a 355 m long basin of trapezoidal cross-section, are more typical of the normal irrigation system layout. Outlet and gravity sluicing control is provided by a double bank of gates. For the Gila basin, the surface loading is about 2.7×10^{-3} m/s at the design discharge of 57 m^3/s. The basin invert gradient is about 0.45%. Further details of the scheme are given in Refs. 2.2 and 7.18.

During the period 1957 to 1972, the All-American works removed an average of 441,000 tonnes of sediment each year, representing a removal efficiency of 61%. The cost of removal has been estimated at $0.40/tonne (1972 prices, (Ref. 7.15)).

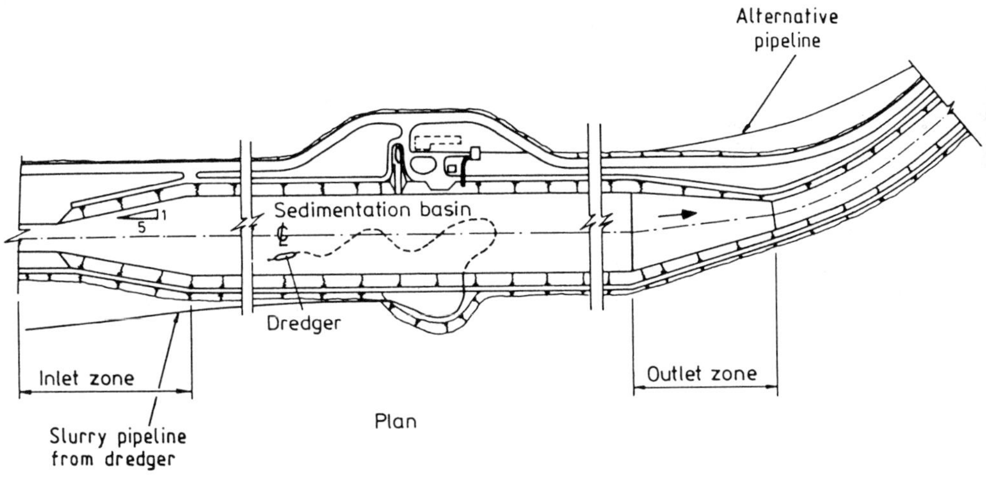

Fig. 7.12 Settling basin with dredger

Fig. 7.12 shows a continuously dredged basin designed for 90% removal of fine sand (0.06 mm) from the head reach of an irrigation canal. The surface loading is 0.95 x 10^{-3} m/s at the design discharge of 40 m^3/s. Flow-through velocity is 0.2 m/s. Two dredgers are provided - the main duty dredger returning the dredged slurry to the parent river, the other acting as standby and able to pump to an alternative disposal lagoon.

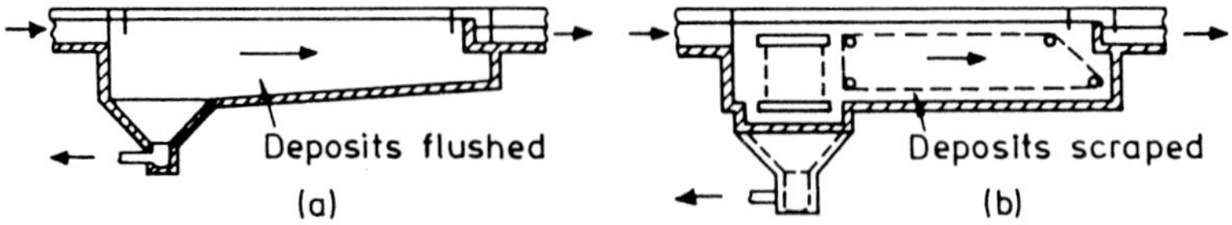

Fig. 7.13 Representative designs of horizontal flow settling tanks

In comparison with irrigation works, Fig. 7.13 shows typical cross-sections of horizontal-flow settling tanks used in water supply systems and Fig. 7.14 shows an enclosed layout used in a hydropower scheme (Ref. 7.19). Other examples are given in Refs. 7.1, 7.11, 7.20 and 7.21. Design surface loading can range from 3.0 x 10^{-5} to 3.0 x 10^{-3} m/s (0.1 to 10 m/hour) depending on the size of particles to be removed.

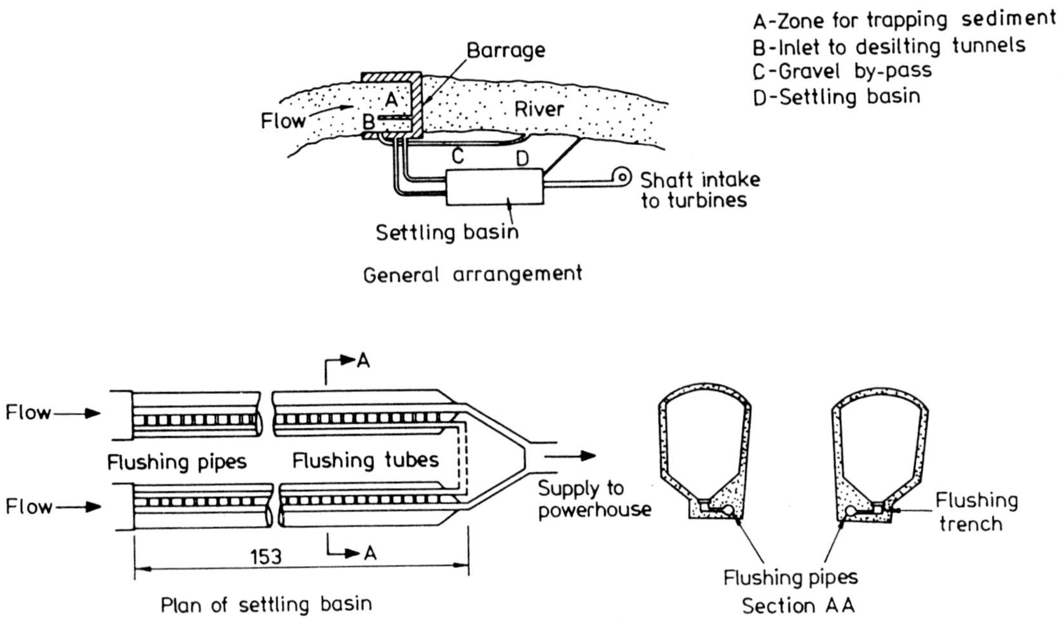

Fig. 7.14 Layout of settling basins in hydropower scheme

7.7 Design Procedure

The suggested design procedure is as follows:

(a) Review estimated suspended sediment inflow rates passing headworks (particle size distribution, concentration, variability).

(b) Review tolerable suspended sediment inflow to system downstream of basin.

(c) Decide on minimum particle size for (say) 90% removal. Estimate basin surface loading using Vetter's equation (equation 7.7) and calculate basin surface area. Subdivide suspended sediment inflow into particle size bands. Estimate removal efficiency for each band, size distribution of sediment deposit, size distribution and concentration of outflow. Compare basin outflow with tolerable inflow to downstream system. Repeat if necessary.

(d) Review local topographic and environmental factors. Consider available head through settling basin system in relation to sediment disposal location and system downstream of basin. Review any relevant local experience with settling basins. Review economics of increased maintenance costs versus degree of sediment removal. Make broad decisions on basin size, method of sediment disposal, extent of standby capacity, and general layout of works.

(e) Review constraints on basin depth and flow-through velocity for preliminary basin layout from considerations of:

- practicality of construction and operation,
- basin turbulence,
- bed scour.

(f) Review, using Camp's solution (Fig. 7.5) and a conservative estimate of turbulence function, the initial estimate of basin area. Optimise basin length and width within constraints of L/W ratio, minimum basin depth and maximum flow-through velocity.

(g) Calculate final design deposition rate, size distribution of sediment deposit, and size distribution and concentration of suspended sediment in outflow as in Step (c).

(h) Estimate additional sediment storage requirement below settling zone based on design deposition rate, method/frequency of sediment removal, and (where appropriate) available sluicing discharge. For gravity sluicing, finalize hydraulic design of sediment removal system. Fix invert levels of basin.

(j) Finalize hydraulic design of inlet and outlet zones. Review need for model test of basin hydraulic layout.

References

7.1 Skeat, W.A. (editor). Manual of British water engineering practice, Institute of Water Engineering Societies, 4th edition, Vol III, Heffer, Cambridge, 1969.

7.2 Tebbutt, T.H.Y. Principles of water quality control, 2nd edition, Pergamon Press, Oxford, 1977.

7.3 Miller, D.G. Sedimentation, a review of published work, Water and Water Engineering, Vol 68, 1964, pp 52-58.

7.4 Hazen, A. On sedimentation, Trans ASCE, Vol LIII, 1904, p63.

7.5 Camp, T.R. Sedimentation and the design of settling tanks, Trans ASCE, Vol 111, 1946, Paper no 2285.

7.6 Dobbins, W.E. Effect of turbulence on sedimentation, Trans ASCE, Vol 109, 1944, p629.

7.7 Ruh-Ming Li and Hsich W. Shen. Solid particle settlement in open-channel flow, Proc ASCE, Vol 101, HY7, July 1975, Paper no 11460.

7.8 Sumer, B.M. Settlement of solid particles in open-channel flow, Proc ASCE, Vol. 103, HY11, November 1977, Paper no 13368.

7.9 Vetter, C.P. Technical aspects of the silt problem on the Colorado River, Civil Engineering, Vol 10, No 11, November 1940, pp 698-701.

7.10 Pemberton, E.L. and Lara, J.M. A procedure to determine sediment deposition in a settling basin, Sedimentation Investigations Technical Guide Series, Section E, Part 2, Bureau of Reclamation, US Department of the Interior, August, 1971.

7.11 Fair, G.M, Geyer, J.C. and Okun, D.A. Water and wastewater engineering, Wiley, New York, 1968.

7.12 Clements, M.S. Velocity variations in rectangular sedimentation tanks, Proc ICE, Vol 34, 1966, p 171.

7.13 Clements, M.S. and Price, G.A. Some lessons from model and full scale tests in rectangular tanks, Water Pollution Control, Vol 73, 1974, No 1.

7.14 Camp, T.R. A study of the rational design of settling tanks, Sewage Works Journal, Vol VIII, No 5, September 1936, p 742.

7.15 International Commission on Irrigation and Drainage, Transactions, 9th Congress, Moscow, 1975. Vol II, Question 30, Sediment-controlling irrigation intake structures.

R9 Hariri, D. and Parhami, F. Desilting structure for the Moghan canal, Iran.

R29 Moser, T.H. and Sears, W.D. Sediment control at Imperial Dam.

R32 Shimura, H. and Kinose, K. Method of improving the performance of a settling basin at a water intake of an alluvial fan river.

R35 Toru, K. Design of an irrigation water intake.

7.16 Ackers, P. Sediment transport problems in irrigation systems design, Chapter 4 - Developments in hydraulic engineering - Volume 1, Novak, P (editor), Applied Science Publishers Ltd., 1983.

7.17 Bagnold, R.A. An approach to the sediment transport problem from general physics, Geological Survey, Professional paper 422-I, US Govt Printing Office, Washington, 1966.

7.18 Dominy, F.E. Design of desilting works for irrigation systems, Proc ASCE, Vol 92, IR4, December 1966, Paper no 4998.

7.19 McKeogh, E.J. Sedimentation control in Himalayan reservoirs, Water Power & Dam Construction, Vol 33, No 10, October 1981, pp 41-43.

7.20 Babbitt, M.S., Doland, J.J. and Cleasby, J.L. Water supply engineering, 6th edition, McGraw-Hill, New York, 1962.

7.21 Twort, A.C., Hoather, R.C. and Law, F.M. Water supply, 2nd edition, Arnold, London, 1974.

Chapter 8

EXAMPLES OF MODIFICATIONS TO INTAKES TO IMPROVE SEDIMENT CONTROL

8. EXAMPLES OF MODIFICATIONS TO INTAKES TO IMPROVE SEDIMENT CONTROL

Many factors which must be considered in the design of intake structures with regard to sediment control including types of intake structures and sediment excluding and controlling devices have been cited in previous chapters. Often, however, the problems associated with sediment ingress into intakes are not appreciated until the system is in use. This chapter gives examples of studies which have been carried out in an attempt to alleviate sediment problems at existing intakes.

(1) The Greater Mussayib irrigation system in Iraq was completed in 1956. Problems of sedimentation and poor distribution were investigated and designs for the remodelling works were completed in 1982 by Binnie and Partners (Ref. 8.1).

The principal maintenance problem lies in the smaller canals where the sediment carrying capacity is less than the main canal. Sedimentation is particularly severe in the head reaches of the branch canals.

It was proposed that the amount of fine sand and coarse silt entering the system would be controlled by:

- remodelling the headworks approach channel geometry so as to abstract preferentially the surface flow in the river (Fig. 8.1).

- provision of a settling basin in the Main canal downstream of the headworks.

Stopping the finer material from entering the canal system would be uneconomic. The head available at the intake was less than 0.5 m - this limited the options for remodelling. No alternative location was available for constructing a new headworks.

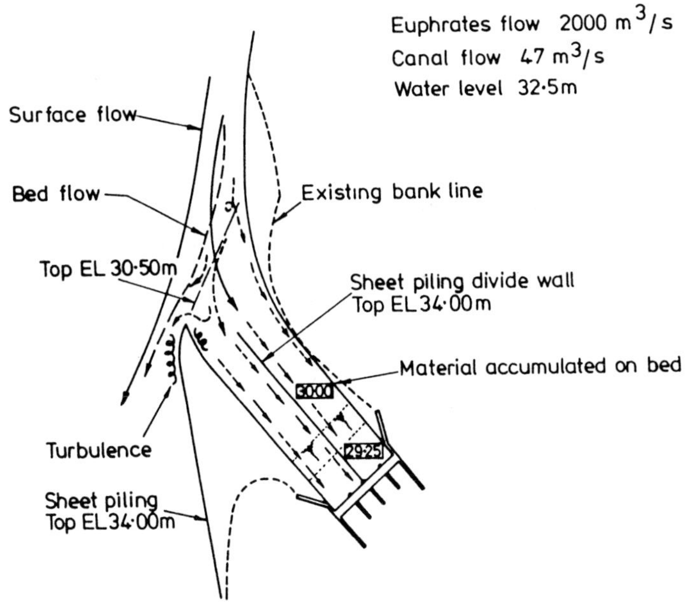

Fig. 8.1 Layout and flow pattern at remodelled headworks

A 1:50 scale model was constructed to examine the performance of the existing headworks and then various modified layouts.

The layout shown in Fig. 8.1 was found to give best all-round performance and was adopted for detailed design. The vertical walls will be constructed mainly in sheet piling. Its principal features are:

- realignment of the upstream river bank to produce a smooth approach alignment with no indentations,
- a submerged sill projecting into the river to train the bedflow away from the approach channel,
- a clearly defined approach channel which maintains the velocity of approach to the head regulator sufficiently high to avoid sediment deposition. The entrance is aligned with the river flow,

- a divide wall to prevent circulation of flow in the approach channel,

- a curved nose at the junction of the downstream bank wall with the sill to reduce local turbulence and scour downstream.

The settling basin was designed using Vetter's equation as a single in-line basin, to remove 90% of fine sand with a main canal flow of 40 m^3/s^{-1}. It is estimated that a maximum volume of 250,000 m^3 of sediment will be removed from the basin annually.

The operation and control of the system with respect to supply and sedimentation has been investigated thoroughly and appropriate recommendations made (Ref. 8.1).

(2) Sediment transport and deposition problems at headworks in Indonesia have been the cause for 65% of the rehabilitation works necessary in hydraulic structures. Ref. 8.2 discusses a number of case histories indicating how problems caused by changes in river regime and sediment transportation have been overcome.

(3) At the Glenfinnan hydro-electric power schemes' Stalker's Intake it was found that the intake chamber was completely filled with moraine which had washed down from the hillside above. The improvement work comprised removal of the original intake structure and blanking off the pipe, construction of an overfall intake chamber that would allow stones and gravel to pass over it without choking, and connection of the new intake chamber to the turbine pipe. In addition, two gabion weirs were constructed at locations upstream of the intake to form gravel traps.

With the new intake arrangement, entrained air was being drawn into the pipeline, resulting in rough running of the Turbine at high flows and, in particular, causing trouble with the domestic water supply through a range of flows. A separate air release chamber adjoining the intake was added to solve this problem.

Fine screen panels were introduced for seasonal attachment over the gravel screen bars to prevent blockage of the intake rose by leaves and grass during the autumn/winter season.

In the first 16 months after recommissioning the upper gabion weir was cleared of gravel once and both weirs were in satisfactory condition having passed some high floods. No gravel choking at the intake had been experienced. The air release chamber was operating satisfactorily.

The improvements are shown in Fig. 8.2.

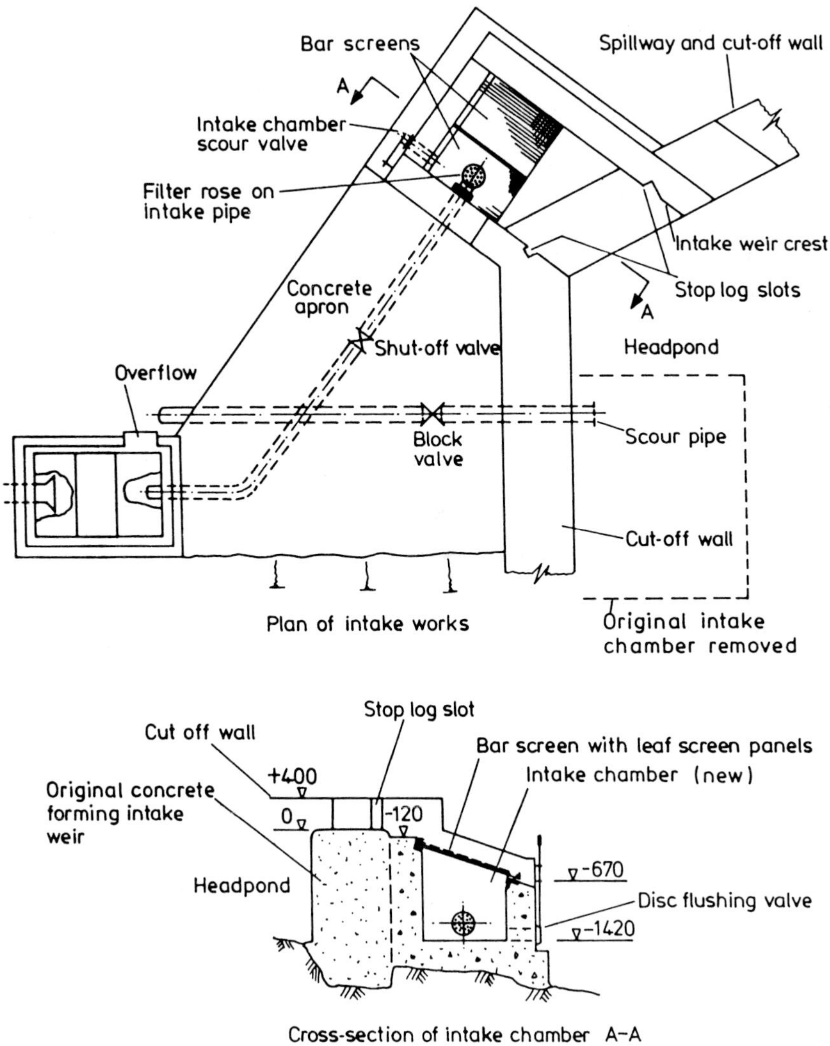

Fig. 8.2 Glenfinnan power schemes - Stalkers intake improvements

(4) Irrigation water for the Ganges-Kobadak Project is withdrawn from the Ganges River about 1 mile downstream from the Hardinge Bridge. Since the start of the project, in the early 60's, problems have been experienced due to the presence of sand banks in front of the intake and excessive sedimentation in the intake canal which necessitates extensive maintaining dredging works.

The sedimentation problems were studied recently (Ref. 8.3), within the framework of a study for the rehabilitation and the improvement of this irrigation project. It was found, from a number of different data, that bars with a width of more than 1 km are traversing through this particular stretch of the Ganges River, and that the quantities of sediment to be dredged yearly could be correlated with this bar movement.

After a careful analysis of the technical and economical aspects of the various possible measures, recommendations were made for the improvement of the present sediment handling procedure by gradually dredging the channel through the sand bar, if any, and for increasing the width and depth of the settling basin to encourage sedimentation and, therefore, to reduce the sediment entering the main irrigation canals.

In addition to this new model tests are recommended to study the effectiveness of floating guide vanes in the Ganges River in front of the intake, taking into account the various river bed configurations which have been observed in the past.

References

8.1 Bramley, M.E. and Thompson, G. "Design of the Remodelled Headworks and Canal System for the Greater Mussayib Project" - Proc 1st Int Conf on Hydraulic Design in Water Resources Engineering : Channels and Channel Control Structures, University of Southampton, April 1984.

8.2 Memed, M. "Some methods to overcome the sediment transport problems on headworks in Indonesia" - Proc 1st Int Conf on Hydraulic Design in Water Resources Engineering : Channels and Channel Control Structures, University of Southampton, April 1984.

8.3 Sarker, A.B., Klaassen, G.J. and Radu, M.Th. "Reducing sediment problems of the Ganges-Kobadak Irrigation Project (Bangladesh) - Proc 1st Inst Conf on Hydraulic Design in Water Resources Engineering : Channels and Channel Control Structures, University of Southampton, April 1984.

APPENDICES

APPENDIX 1 : DESCRIPTIONS OF TYPICAL INTAKE FEATURES

Forebay Open area of water upstream of the intake structure.

Floating Boom Situated upstream of or in the forebay for diverting flotsam.

Weir Low, overflow type control structure constructed across the river to divert flow into the intake.

Barrage Gated control structure.

Headpond Pond formed by river weir or excavated or a reservoir to provide head at the intake - can also provide primary sedimentation.

Divide Groin/Wall Runs upstream from the weir or barrage to separate flows in the approach channel.

Inlet or Entry Sill A step across the face of the intake structure or settling basin to permit abstraction of less heavily sediment laden near-surface flow.

Skimmer Wall A hanging wall across the face of the intake structure to prevent ingress of flotsam.

Splitter Piers Form a series of narrow openings in the face of the intake structure so that large flotsam is excluded from the face of the screens (- also provide ability to shut off individual pump supplies for maintenance if walls connect piers to sump walls).

Coarse Screen or Trash Rack Set in the face of the intake structure to exclude surface and sub-surface trash, frazil ice, etc. usually with provision for clearing by raking.

Sluice Gate Generally behind the coarse screen in the inlet to the intake structure, for closing off the structure from the source, i.e. river or reservoir.

Under Sluice, Scour or Bottom Outlet A low level channel controlled by a sluice gate for passing bedload from the headpond through the weir, usually located adjacent to the intake sill. Under sluice also passes near-bed flow with higher suspended sediment concentration.

Settling Basin or Sand Trap A basin incorporated in or close to the intake structure, where low velocity causes some or all of the sediment load in suspension to settle out of the flow.

De-Silting Channel or Canal Channel by which deposited sediment can be scoured out.

De-Silting Sluice Gate Controls flow to the de-silting channel.

Transition Gradually varying channel cross-section within the intake structure, to control the inlet velocity and maintain streamline flow in the waterways associated with the intake.

Intake Gate or Regulator A gate capable of operating against unbalanced head that can be used for shutting off and controlling the flow from an intake structure into the conduit.

Conduit or Channel Pipe, tunnel or canal for conveying the supply of water from the intake.

APPENDIX 2 : SEDIMENT TRANSPORT THEORY

A2.1 Scope

There is a very large quantity of literature on sediment transport theory and related issues such as alluvial channel resistance and river and canal regime. The scope of this Appendix is limited to a summary of those aspects most relevant to sediment control at intakes, selecting a few of the many available formulations that will be of value in considering the general transport of bed materials and solids in suspension in the river, through the intake works and hence the distribution system. Local influences generated by the three-dimensional nature of the flow at the river/intake interface are not covered. The situations referred to in the basic theories of sediment transport are with steady uniform flow in straight channels.

A2.2 Terminology and Definitions

Recent International and British Standards (Ref. A2.1) are available and the following extracts from BS 3680 10B (1980) provide the accepted terminology and definitions, illustrated also in Fig. A2.1.

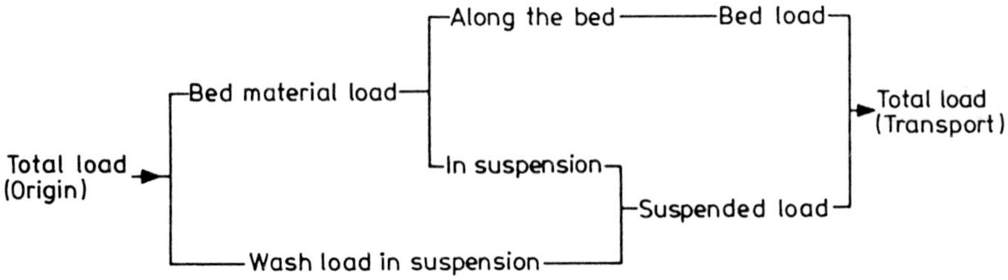

Fig. A2.1 Modes and definitions of sediment transport

"For a proper comprehension of sediment movement and related terms, the flow of water over an artificially flattened bed of sediment may be considered. From no movement of bed material at very low velocities, some particles begin to move with the increase of velocity by sliding, rolling or hopping along the bed (bed load); at still higher velocities particles from the bed are thrown into suspension by turbulence (suspended load). The suspended load also includes finer particles in near permanent suspension brought in from the catchment (wash load). Bed load and suspended load may occur simultaneously, but the borderline between them is not well defined.

Total load : From the view point of transport of sediment, the total load comprises bed load and suspended load, the latter including wash load. From the view point of origin of the sediment, the total load comprises the bed material load (including the suspended portion) and the wash load (see Fig. A2.1).

Bed material : The material, the particle sizes of which are found in appreciable quantities in that part of the bed affected by transport.

Bed material load : The part of the total sediment transport which consists of the bed material and whose rate of movement is governed by the transporting capacity of the channel.

Suspended load : That part of the total sediment transported which is maintained in suspension by turbulence in the flowing water for considerable periods of time without contact with the stream bed. It moves with practically the same velocity as that of the flowing water. It is generally expresses in mass or volume per unit of time.

Bed load : The sediment in almost continuous contact with the bed, carried forward by rolling, sliding or hopping.

Wash load : That part of the suspended load which is composed of particle sizes smaller than those found in appreciable quantities in the bed material; it is in near permanent suspension and, therefore, is transported through the stream without deposition. The discharge of the wash load through a reach depends only on the rate with which these particles become available in the catchment and not on the transport capacity of flow. It is generally expressed in mass or volume per unit of time."

A2.3 Fundamentals of Bed Material Movement

The basic quantities which influence the process of sediment transport in two-dimensional, free surface flow are the unit mass of fluid ρ, the unit mass of solids ρ_s, kinematic viscosity of the fluid ν, particle diameter D, water depth d, shear velocity $\sqrt{(\tau_o/\rho)}$ or $\sqrt{(gdS)}$, denoted v_* and acceleration due to gravity g. Dimensional analysis yields the following grouping of these basic quantities:

Dimensionless grain size

$$D_{gr} = \left\{ \frac{\gamma_s D^3}{\rho \nu^2} \right\}^{1/3} = \left\{ \frac{g(s-1)}{\nu^2} \right\}^{1/3} \times D \qquad A.1$$

Mobility number

$$Y = \frac{\rho v_*^2}{\gamma_s D} = \frac{v_*^2}{(s-1)\, gD} \qquad A.2$$

Relative grain size or dimensionless flow depth

$$Z = \frac{d}{D} \qquad A.3$$

Relative mass density, solid/liquid

$$s = \frac{\rho_s}{\rho} \qquad A.4$$

Hence, any property related to the movement of bed material in steady, uniform, two-dimensional flow is a function of these four dimensionless groups, in particular, the dimensionless sediment transport parameter.

$$\frac{q_t \rho^{1/2}}{\gamma_s^{3/2}\, D^{3/2}} = f\,(D_{gr}, Y, Z, s) \qquad A.5$$

where the left-hand side is the Einstein (Ref. A2.2) transport function ϕ, q_t is the sediment transport rate as submerged weight per unit width per unit time and γ_s is the submerged unit weight of the solid phase.

$$\phi = \frac{q_t \rho^{1/2}}{\gamma_s^{3/2} D^{3/2}} = \frac{q_t}{(s-1)^{3/2} g^{3/2} \rho D^{3/2}} \qquad \text{A.6}$$

These parameters, D_{gr}, Y, Z, s and Ø are found in various combinations in many theories of sediment transport.

There is a special case of the general function in eqn A.5, namely the condition at which bed material is just on the point of movement with $q_t = 0$.

$$0 = f\ (D_{gr},\ Y_{cr},\ Z,\ s) \qquad \text{A.7}$$

In practice, the relative density is fully accounted for in the parameters D_{gr} and Y, and so may be omitted. This 'initial motion' condition is then usually expressed in the form

$$Y_{cr} = f_o\ (Y_{cr}^{1/2}\ D_{gr}^{3/2}\ ,\ Z) \qquad \text{A.8}$$

The ratio, Z, of depth to sediment diameter is of very minor significance except in relatively shallow flow, and so equation A.8 simplifies to the Shields function (Ref. A2.3).

$$\frac{v_*^2}{(s-1)\ g\ D} = f_o\ (\frac{v_*\ D}{\nu}) \qquad \text{A.9}$$

This is shown in Fig. A2.2 with various sources of data and an additional set of grid lines related to constant sediment properties.

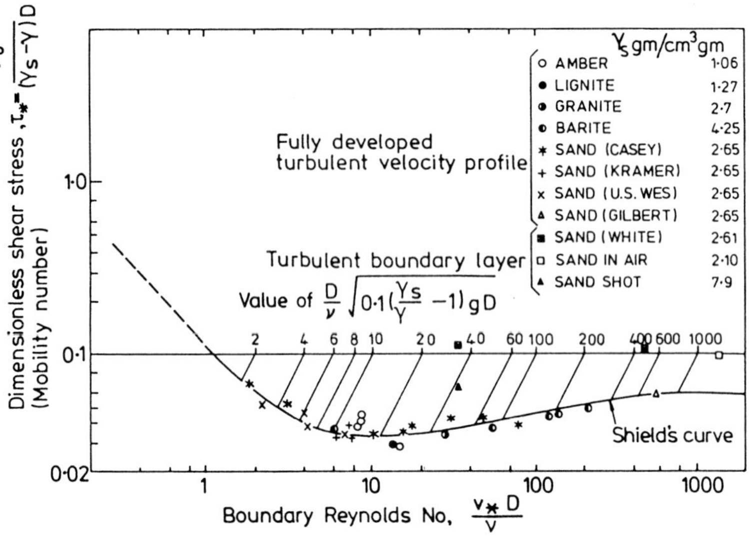

Fig. A2.2 Shields function for incipient motion

When Y exceeds Y_{cr}, then sediment transport becomes established and the transport function Ø increases progressively (i.e. q_t increases) as the power of the stream increases diagonally upwards from the initial motion condition. The assessment of sediment transport thus requires knowledge of how Ø varies from zero as one moves upwards and to the right in Fig. A2.2.

A2.4 Available Formulae for Bed Material Load

Ref. A2.4 reviewed available formulae (as in 1975) in the context of laboratory and field data and concluded that few were of acceptable accuracy. The best were the methods of Ackers and White (Ref. A2.5) and of Engelund and Hansen (Ref. A2.6), the former being slightly more accurate and the latter being the simpler. Only these two formulations will be quoted, but other methods will be mentioned later. These include a purely empirical method for sand bed rivers and procedures whereby field information on material in suspension may be used to assess the unmeasured transport close to the bed.

A2.5 Ackers and White Method

A2.5.1 Theoretical basis:

This theory first considered coarse sediment and fine sediment separately, and then sought a transitional function between them. The transitional sizes include the sands and silts of great practical interest in rivers and irrigation systems.

A coarse sediment is transported mainly as a bed process. If bed features exist, it is assumed that the effective shear stress bears a similar relationship to mean stream velocity as with a plane grain-textured surface at rest.

A fine sediment is transported within the body of the flow, where it is suspended by the stream turbulence. As the intensity of turbulence is dependent on the total energy degradation, rather than on a net grain resistance, for fine grained material the total bed shear stress is effective in causing sediment transport.

Sediment mobility is described by the ratio of the effective shear force on unit area of the bed to the immersed weight of a layer of grains. This mobility number is denoted F_{gr}, and a general definition is:

$$F_{gr} = \frac{v_*^n}{\sqrt{\{gD(s-1)\}}} \left\{\frac{v}{\sqrt{32}\,\log\,(\alpha d/D)}\right\}^{1-n} \qquad \text{A.10}$$

For coarse sediments (n = 0), the expression reduces to the form

$$F_{gr} = F_{cg} = \frac{v}{\sqrt{\{gD(s-1)}} \quad \frac{1}{\sqrt{32}\,\log\,(\alpha d/D)} \qquad \text{A.11}$$

The coefficient α relates the roughness, k_s, in the rough-turbulent formulation for channel resistance to the median sediment diameter, D. For fine sediments (n = 1);

$$F_{gr} = F_{fg} = \frac{v_*}{\sqrt{\{gD(s-1)\}}} = \sqrt{Y} \qquad \text{A.12}$$

The transition parameter n was evaluated by analysing flume data for a range of sediment sizes and was expected to lie between 0 and 1, showing a continuous variation through coarse material and transitional sizes to fine non-cohesive sand or silt.

Sediment transport is based on the stream power concept; in the case of coarse sediments the product of net grain shear and stream velocity is used as the power per unit area of bed, and for fine sediments the total stream power is used. The useful work done in sediment transport in the two cases takes account of the different modes of transport assumed, and in relation to the stream power gives an expression for the efficiency of the the transport process. Efficiency is expected to be dependent on the mobility number, F_{gr}. Clearly there would be a value of F_{gr} below which no sediment would move and efficiency would be zero. As F_{gr} rises above the limiting value, A_{gr}, it is expected that the efficiency would increase.

In order to separate the primary variables, the efficiency (which is dimensionless) was combined with the mobility number, F_{gr}, to yield a general transport parameter:

$$G_{gr} = \frac{Xd}{sD} \left(\frac{V_*}{V}\right)^n \qquad \text{A.13}$$

where

$$X = g_s/\rho g v d \qquad \text{A.14}$$

X is the mass rate of sediment transport per unit width expressed as a ratio of the mass rate of fluid flux per unit width, which it is convenient to think of as a special form of sediment concentration, g_s is the weight of solids in motion, per unit time.

The general transport function tested was expressed as:

$$G_{gr} = f_1 \, (F_{gr}; D_{gr}) \qquad \text{A.15}$$

i.e. sediment transport is a function of sediment mobility and grain size. As the definitions of G_{gr} and F_{gr} depend on the transition parameter, n, it was also necessary to confirm that

$$n = f_2 \, (D_{gr}) \qquad \text{A.16}$$

Referring back to the functional relationship of eqn A.5, it will be seen that this framework is a special case, reducing the five-variable form to a three-variable form which was more amenable to analysis.

The type of relationship suggested was a power function of G_{gr} with (F_{gr} - A_{gr}), where A_{gr} is the value of F_{gr} at which motion first starts.

$$G_{gr} = J \left(\frac{F_{gr}}{A_{gr}} - 1\right)^{m'} \qquad \text{A.17}$$

A2.5.2 Calibration and Evaluation

Using very many sources of data, a computer-based optimisation routine was developed to obtain best-fit values for the four empirical parameters, J, A_{gr}, m' and n. The variation of these with the dimensionless grain-size D_{gr}, was determined and algebraic functions were fitted to the data for computational convenience.

For transitional sizes, $1 < D_{gr} < 60$:

$$n = 1.00 - 0.56 \log D_{gr} \qquad \text{A.18}$$

$$A_{gr} = \frac{0.23}{\sqrt{D_{gr}}} + 0.14 \qquad \text{A.19}$$

$$m' = \frac{9.66}{D_{gr}} + 1.34 \qquad \text{A.20}$$

$$\log J = 2.86 \log D_{gr} - (\log D_{gr})^2 - 3.53 \qquad \text{A.21}$$

For coarse sediments, $D_{gr} \geq 60$:

$$n = 0.00 \qquad \text{A.22}$$

$$A_{gr} = 0.17 \qquad \text{A.23}$$

$$m' = 1.50 \qquad \text{A.24}$$

$$J = 0.025 \qquad \text{A.25}$$

A round figure value for α of 10 was estimated, not significantly different from the value of 12.3 with the roughness defined by median grain size in the rough-turbulent equation. The transport function is plotted in Fig. A2.3.

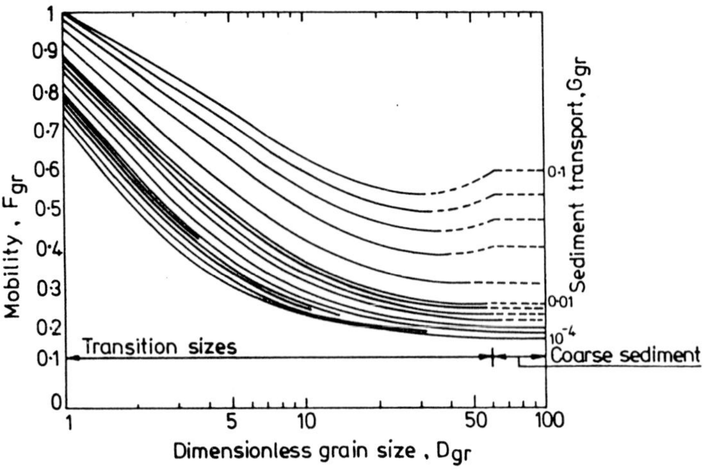

Fig. A2.3 Ackers and White sediment transport function

A2.5.3 Application of Method

To calculate the sediment flux the following properties of the system must be known: (1) Particle size, D; (2) relative density of sediment, s; (3) mean velocity of flow, V; (4) shear velocity, v_*, based on measurements of the velocity distribution or the depth/slope relationship ($v_* = \sqrt{(gdS)}$); (5) depth of flow, d; (6) kinematic viscosity of fluid, ν; and (7) acceleration due to gravity, g.

The calculation proceeds as follows:

1. Determine the value of D_{gr} from known values of D, g, s and ν (equation A.1).

2. Determine the values of n, A, m' and J associated with the derived D_{gr} value (equation A.18 - A.25).

3. Compute the value of particle mobility, F_{gr} (equation A.10).

4. Determine the value of G_{gr} from equation A.17 or from Fig. A2.3 which represents a graphical version of the new sediment transport function.

5. Convert G_{gr} to sediment flux X (the effective mass concentration in the fluid flux), using equation A.13, and the weight rate of transport using equation A.14.

A2.5.4 Graded Sediments

Although the theory was originally checked against flume data for sediments with narrow gradings using the D_{50} size, it has since been compared with the data from the laboratory and the field for graded sediments. For sediments with only a modest range of particle sizes, say $D_{84}/D_{16} < 5$, it was found that the total transport of bed material could be related to the 35 percentile and consequently the particle size D should be interpreted as D_{35} for these applications.

With widely graded sediments it is possible to follow the procedure recommended by Einstein (Ref. A2.7) and others, in which the bed material grading curve is used to consider, say, ten size fractions separately. In this case, however, the assumed threshold conditions for each size fraction should take into account the shielding of the finer fractions and the additional exposure of the coarse material when surrounded by these finer fractions. The calculation procedure in this case is as follows:

1. Consider first the coarsest fraction; from the upper and lower bounds of the fraction determine the mean diameter of the fraction; from the grading curve determine the percentage of the bed material sample in the range.

2. Follow steps 1-5 of the basic procedure given above except that at step 2 a modified value of A, say A^1, should be taken. A^1 is given by the expression $A^1/A = (D_i/D_{50})^{-0.2}$.

3. Factor the transport of this fraction by the sample percentage/100.

4. Proceed to the next coarsest fraction, and so on to the finest. Note that any material below 0.04 mm in the bed sediment grading curve is to be excluded, because it will travel as wash load to which the method does not apply.

5. Total the factored transport rates for all fractions. This is the total estimated bed material load.

A2.6 Engelund and Hansen method

In terms of goodness of fit to laboratory and field data, there is not a great deal to choose between the Engelund and Hansen method (Ref. A2.6) and the Ackers and White method 1 (Ref. A2.5), the latter being marginally superior but more complicated algebraically.

Even the best of the currently available theories are only able to assess transport rates from the hydraulic parameters of a channel to a two-fold accuracy for about two-thirds of the data, however. This disappointing but nevertheless realistic conclusion arises from scatter in the data and from the sensitivity of transport rate to the hydraulic variables, as well as from imperfections of the theories.

The Engelund and Hansen formulation is remarkably simple:

$$\frac{\lambda}{4} \Phi = 0.1Y^{5/2} \qquad \text{A.26}$$

where

$$\lambda = 8gdS/V^2 \qquad \text{A.27}$$

(Φ and Y are defined by eqns A.6 and A.2 respectively)

In their original paper, Engelund and Hansen gave a graphical solution which facilitated the determination of both the water flow and sediment transport rates. The function is dimensionally homogeneous, and utilises the median grain size D_{50} as the characteristic sediment diameter. It has a tendency to overestimate transport at low shear rates, but performs well with laboratory flume data and field results.

A2.7 Colby method

The method proposed by Colby (Ref. A2.8) was not included in the comparisons made by White, Milli and Crabbe (Ref. A2.4), because it consists of empirical sediment transport curves for sands rather than formulae capable of numerical testing. It is nevertheless a basically sound method of assessing the bed material load in sand bed rivers and canals, with median grain diameters of between 0.1 and 0.8 mm, depths from 0.3 m and velocities up to 3 m/s.

Colby's method consists of the simple use of curves based on the correlation of sediment transport data obtained from field measurement. These prediction curves are shown in Fig. A2.4, and provide the bed material load in kg/day per metre width for known values of depth and mean stream velocity. The application of this method is illustrated in section 6.1.

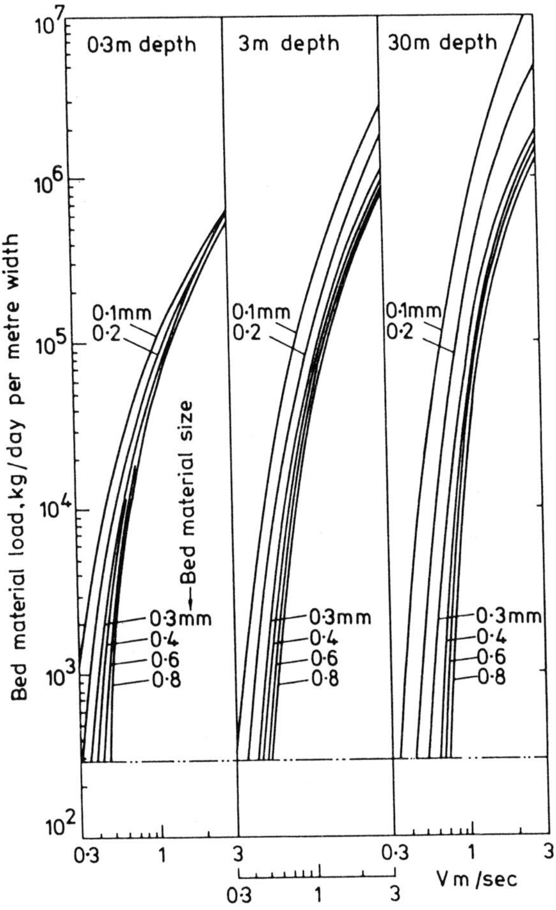

Fig. A2.4 Colby's curves for sand transport

A2.8 Transport in Suspension

In a steady uniform flow carrying sediment in suspension, under equilibrium conditions the concentration at any level will be constant and the net vertical flux at any level will be zero. The settling out of sediment due to the weight of the particles will be balanced by the upward flux due to turbulent diffusion from lower layers with a higher concentration.

The settling rate per unit plan area is Cw where C is the local concentration and w is the terminal velocity of the individual particles. Thus

$$Cw + \varepsilon \frac{dC}{dy} = 0 \qquad \text{A.28}$$

where the second term is the vertical diffusive transport due to the concentration gradient dC/dy, ε being the diffusion coefficient for sediment.

Rouse (Ref. A2.9) developed the above concept by combining it with the basic theory of turbulent flow in an open channel. The assumption was made that the diffusion coefficient for sediment is proportional to the momentum diffusion coefficient in the mixing-length theory of turbulent flow. Utilising also the corresponding flow parameters, e.g. the bed shear stress and overall resistance of the fluid phase, Rouse derived the following function for the vertical distribution of suspended material.

$$\frac{C}{C_a} = \left\{ \frac{(d-y)\ a}{y\ (d-a)} \right\}^z \qquad \text{A.29}$$

where the Rouse number z, is given by:

$$z = \frac{w}{\beta K v_*} \qquad \text{A.30}$$

In the above:

C = concentration at elevation y above the bed
C_a = concentration at a reference elevation a above the bed
d = flow depth
w = fall velocity
v_* = shear velocity (= $\sqrt{gdS}$)
K = von Karman constant (0.4)

β = ratio of sediment diffusion coefficient to momentum coefficient (assumed 1.0)

This function is given in Fig. A2.5. This is taken from Ref. A2.10, which provides further information on the subject.

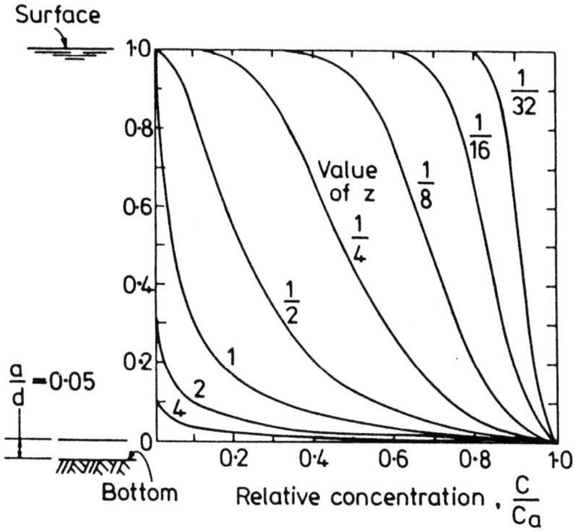

Fig. A2.5 Graph of Rouse suspended load distribution equation for a/d = 0.05 and several values of z

Crucial to any consideration of sediment behaviour is a knowledge of the fall velocity of discrete particles. This may be determined from the Rubey equation (Ref. A2.11) or a graphical presentation as in Fig. A2.6 (from Ref. A2.12). The Rubey equation is for natural particles:

$$w = \sqrt{gd\ (s-1)}\ \left\{\left[\frac{2}{3} + \frac{36\nu^2}{gd^3\ (s-1)}\right]^{\frac{1}{2}} - \left[\frac{36\nu^2}{gd^3\ (s-1)}\right]^{\frac{1}{2}}\right\} \qquad A.31$$

In the above, ν is the kinematic viscosity of the fluid which is a function of water temperature, as follows:

Temp ^oC :	0	5	10	15	20	25	30
ν, $10^{-6}m^2/s$:	1.79	1.53	1.32	1.15	1.02	0.91	0.81

Fig. A2.6 includes the influence of shape, with the shape factor designed as SF = $a/\sqrt{(bc)}$ where a, b, and c are the mutually perpendicular dimensions (a being the smallest).

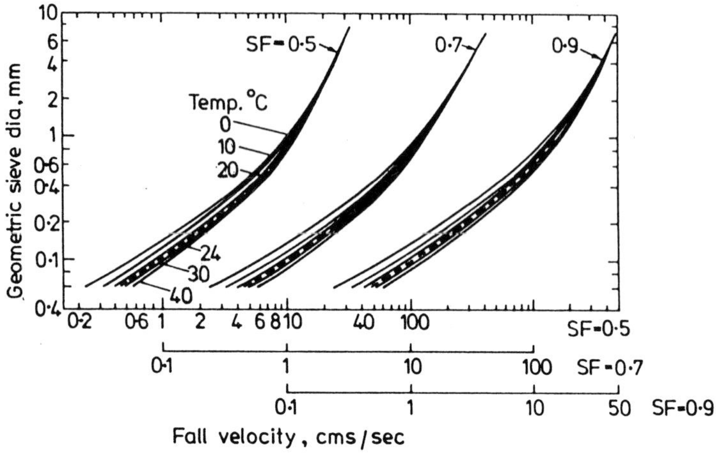

Fig. A2.6 Relation between sieve diameter, fall velocity and shape factor (SF) for naturally worn quartz sand particles falling in distilled water

A2.9 Calculation of Sediment Discharge from Stream Measurements

The development of the depth-integrating sampler (see Chapter 2 and Section A Ref. A2.10) made it possible to sample the suspended load of streams on a routine basis. In many countries there are national agencies responsible for collecting data of this kind as well as hydraulic data on streams but where data is not already available, it may be collected specifically for a project using standard methods (e.g. Ref. A2.1). Suspended load samplers cannot sample the flow within a layer several centimetres thick at the bed and do not include any of the bed load. For fine sediments, e.g. those making up the wash load, which are nearly uniformly distributed over the depth, the concentrations measured are essentially equal to the discharge concentration. However, for the bed sediment that is coarser than the wash load, the concentrations in the samples are usually considerably less that the true mean concentration in the flow. To get the true sediment discharge it is necessary to add the contribution of the unsampled suspended load and the bed load to the measured sediment discharge.

Methods of estimating the suspended-sediment discharge in the unsampled layer near the bed and the bed load discharge have been proposed. The Colby and Hembree method (Ref. A2.13) is known as the Modified Einstein Method and is based on observed suspended load samples, mean velocity, depth, cross-section, and size composition of the bed sediment. Its name derives from the fact that it uses a modified version of the Einstein bed load function (Ref. A2.14) in estimating the unmeasured sediment discharge. Sediment discharge data based on suspended-sediment samples and estimates of the unmeasured discharge are much more reliable than those based on the formulae alone. The methods of estimating unmeasured sediment discharge are presented in detail in Ref. A2.10.

A2.10 Sediment Transport Through Lined Channels, Pipes and Culverts

Most treatments of sediment transport in the civil engineering context relate to channels with an alluvial bed. Bed material transport is then not restricted by supply : if the local stream power exceeds that needed to convey the incoming sediment load, the stream will pick up extra material from the bed to achieve a new balance. Over the concrete bed of an artificial channel or within a pipeline, this balance does not exist if there is no deposition to form a cover of alluvium over the solid boundary. Several stages of sediment transport may occur:

1. Supply less than the transporting capacity of the flow over a clean solid surface.

2. Supply just in balance with the transport capacity so there is incipient deposition.

3. Supply exceeding the incipient condition so that deposition occurs. This deposition, through changing the available cross-section and increasing roughness, will alter the flow conditions and may reduce the sediment transport capacity below that of stage 2 and so deposition may accelerate.

4. Transport capacity over the deposited bed in balance with the supply.

Condition 2 above is of crucial importance. Where the flow and/or sediment supply varies in time, the system has either to be designed securely in the range of condition 1 so that the lined channel, pipe or culvert remains permanently free of deposition; or it should be designed for condition 4, with allowance made for the influence of deposition on the capacity of the system to transport both fluid and sediment. Conventional sediment transport theory and methods apply to the latter zone, but special equations apply to the former zone, in effect providing the limiting condition under which the maximum supply of sediment may be conveyed without deposition over a solid surface.

The subject of pipe-line transport is reviewed in Ref. A2.10, and also by Graf (Ref. A2.15) who has himself developed the limiting function

$$\frac{C_v V_s R}{\sqrt{(s-1) g D^3}} = 10.39 \left[\frac{(s-1) D}{SR}\right] \qquad \text{A.32}$$

where C_v is the volumetric sediment concentration, R is the hydraulic radius of the flow section and V_s is the flow velocity at incipient deposition.

A more recent treatment of the subject by Novak and Nalluri (Ref. A2.16) yields as the limiting condition for deposition:

$$S = (s-1) \left(\frac{D}{R}\right)^{1/3} \left\{\frac{C_v R^{1/6}}{11.6\, n_m \sqrt{g}}\right\}^{2/3} \qquad \text{A.33}$$

Laboratory experiments by May (Ref. A2.17) gave the limiting velocity V_s as a function of C_v, for pipes of diameter D_o.

$$C_v = 0.0205 \left(\frac{D_o}{A}\right)^2 \left(\frac{D}{R}\right)^{0.6} \left\{\frac{V_s^2}{g\,(s-1)\,D_o}\right\}^{3/2} \left\{1 - \frac{V_o}{V_s}\right\}^{-4} \qquad \text{A.34}$$

in which V_o, the limiting condition for zero transport, is given by Novak and Nalluri's formula

$$V_o = 0.61 \sqrt{gD\,(s-1)} \left(\frac{D}{R}\right)^{-0.27} \qquad \text{A.35}$$

These various methods give rather different results, especially when applied to large systems well beyond the range of experimental data. Designing for operation without any deposition should recognise the uncertainty in the methodology and so include something in reserve.

References

A2.1 British Standards. Methods of measurement of liquid flow in open channels BS 3680, Part 1, 1983, Glossary of terms; BS 3680, Part 10, Sediment transport; Part 10B, 1980, Measurement of suspended sediment; Part 10C, 1980, Bed material sampling.

A2.2 Einstein, H.A. The bed load for sediment transportation in open channel flow, Ser Tech Bull US Dept Agric Soil Conserv, 1950, No 1026.

A2.3 Shields, A. "Anwendung der Aenlichkeitsmechanik und der Turbulenzforschung auf die Geschiebebewegung", Mitteilungen der Preussischen Versuchsanstalt fur Wasserbau und Schiffbau, Berlin, Germany, translated to English by W P Ott and J C van Uchelen, California Institute of Technology, Pasadena, California, 1936.

A2.4 White, W.R., Milli, H. and Crabbe, A.D. Sediment transport theories, a review, Proc Instn Civil Engineers, Part 2, June 1975, 59, 265-92.

A2.5 Ackers, P. and White, W.R. Sediment transport: new approach and analysis, Proc ASCE, November 1973, 99 (HY11), 2041-2060.

A2.6 Engelund, F. and Hansen, E. A Monograph on Sediment Transport in Alluvial Streams, Teknisk Vorlag, Copenhagen, 1967.

A2.7 Einstein, H.A. Bed load transportation in Mountain Creek, US Dept of Agriculture, Washington DC, 1944.

A2.8 Colby, B.R. Discharge of sands and mean velocity relationships in sand-bed streams, US Geological Survey, Professional Paper 462-A, US Government Printing Office, Washington, 1964.

A2.9 Rouse, H. Modern conceptions of the mechanics of fluid turbulence, Trans ASCE, Vol 102, paper 1965, 1937, pp 463-543.

A2.10 Vanoni, V.A. (ed), Sedimentation engineering, ASCE Manuals and reports on engineering practice no 54, ASCE, 1975.

A2.11 Rubey, N.W. Settling velocities of gravel, sand and silt particles, Amer J of Science, 5th series, vol 25, no 148, 1933, p 325-338.

A2.12 Interagency Committee. Some fundamentals of particle size analysis, a study of methods used in measurement and analysis of sediment loads in streams, Report No 12, Subcommittee on sedimentation, Interagency committee on water resources, St Anthony Falls Hydraulic Laboratory, Minneapolis, Minn, USA, 1957.

A2.13 Colby, B.R. and Hembree, C.H. Computations of total sediment discharge, Niobrara River near Cody, Nebraska, Water Supply Paper 1357, US Geological Survey, Washington DC, 1955.

A2.14 Einstein, H.A. The bed load function for sediment transportation in open channels, Technical Bulletin 1026, US Dept of Agriculture, Soil Conservation Service, Washington DC, 1950.

A2.15 Graf, W.H. Hydraulics of sediment transport, McGraw-Hill, New York, 1971.

A2.16 Novak, P. and Nalluri, C. Sewer design for no sediment deposition, Proc Instn Civil Engineers, Part 2, Vol 65, September 1978, p 669-674.

A2.17 May, R.W.P. Sediment transport in sewers, Report IT222, Hydraulics Research Station, Wallingford, February 1982.

APPENDIX 3 : EXAMPLE CALCULATION

For the purpose of this calculation a converging curved channel excluder is assumed, with a layout generally as shown in Fig. 6.1.2.

Data

Canal Flow: 20 m^3/s

Sluice Flow: 7 m^3/s at low flow conditions and average water availability

River Bed Sediment: A median diameter of 0.5 mm is assumed with a typical size distribution

Layout

Main dimensions have to be assumed, and the design checked.

(a) Depth of flow: For low flow conditions the canal headgates will be open and it may be assumed that the depth of flow is controlled by flow through the canal head regulator.

The discharge intensity, q, at the head regulator is 20/(4 x 5), i.e. 1 $m^3/s/m$. The minimum upstream level in the sluiceway will be determined by the head required to pass this discharge through the canal head regulator. Assuming this is equivalent to the discharge over a broad-crested weir.

i.e. $$q = 1.706\ H^{3/2}$$

$$H = (1/1.706)^{2/3} = 0.70\ m$$

For a skimming weir height of 0.5 m, depth of flow in the curved channel will be approximately 1.2 m. This would have to be maintained by controlling the sluice gate opening.

(b) Assumed inner curve radius = 20.85 m (r_i)

Assumed outer curve radius = 29.15 m (r_o)

$$Q/d = V_i r_i \ln (r_o/r_i)$$

Opposite central headgate pier, Q = 17.0 m^3/s approximately.

$$\frac{17.0}{1.2} = V_i \times 20.85 \ln (29.15/20.85)$$

$$V_i = 2.03 \text{ m/s}$$

$$V_o = 2.03 \times \frac{20.85}{29.15} = 1.45 \text{ m/s}$$

(c) The average velocity in the curved channel at the upstream entrance to the canal head regulator is approximately

$$V = 27/(1.2 \times 8.3) = 2.71 \text{ m/s}$$

$$\text{Froude number Fr} = 2.71 \sqrt{1.2g} = 0.79$$

CRITICAL SHEAR STRESS APPROACH

To calculate mean velocity corresponding to critical shear stress, use equation 6.1.2.

$$I = \frac{D}{\nu} (0.1 (\gamma_s/\gamma - 1) g d_s)^{\frac{1}{2}}$$

$$= \frac{0.53 \times 10^{-3}}{0.73 \times 10^{-6}} (0.1 \times 1.65 \times 9.81 \times 0.5 \times 10^{-3})^{\frac{1}{2}}$$

$$I = 19.49$$

Then from Fig. A2 (and equation 6.1.3)

$$\tau_* = \frac{v_*^2}{g D (\gamma_s/\gamma - 1)} = 0.032$$

Hence,

$$v_*^2 = 0.032 \times 9.81 \times 0.5 \times 10^{-3} \times 1.65$$

$v_* = 0.016$ m/s

Then,

$$V = 2.5\ v_*\ \ln \frac{12.3\ R}{\Delta} \qquad 6.1.4$$

To establish value of effective roughness, Δ:

The reference laminar sublayer thickness, $\delta = \frac{11.6\ \nu}{v_*}$ (from roughness theory)

so here $\delta = 0.5 \times 10^{-3}$ m

Supposing the roughness height k of the sand to be 0.57 mm

$$k/\delta = \frac{0.57}{0.5} = 1.14$$

From a graph of effective roughness coefficient, x, against k/δ ,

$x = 1.6$

Effecive roughness, $\Delta = \frac{0.57 \times 10^{-3}}{1.6} = 0.36 \times 10^{-3}$ m

From equation 6.1.4, where $R = (8.3 \times 1.2)/(8.3 + (2 \times 1.2)) = 0.93$ m

$V = 0.41$ m/s

From this calculation it is clear that the actual velocity in the curved sluice channel is far in excess of that required to produce critical shear conditions on a flat bed of 0.5 mm sand.

Check corresponding to (b) above on the section immediately adjacent to the sluice gates:

Approximately, discharge = 7 m^3/s

Average velocity = 7/(1.2 x 8.3) = 0.7 m/s

It is seen that the actual velocity is about 1.7 times that corresponding to critical shear stress on a flat bed of 0.5 mm sand.

A value of τ_* can be computed based on an average velocity of 0.7 m/s, a depth of flow of 1.2 m and a roughness height k = 0.57 mm; R = 0.93 m.

A value of v_* = 0.027 m/s is calculated giving

τ_* = 0.09

Of course, τ_* will vary throughout the sluice channel and this is only one representative point. More investigation is required for this criterion.

CHECK BY COLBY'S SEDIMENT TRANSPORT CURVES (FIG A2.4)

Since the excluder in Fig. 6.1.2 was model tested and found to work quite satisfactorily, it seems that the critical shear stress will always be exceeded by a certain factor. In fact, there may be some build up of sediment in the sluice channel so that the bed is not flat. An alternative approach is to check the sediment transporting power of the sluice flow. By interpolation from Fig. A2.4, for a flow depth of 0.82 m and a mean velocity of 1.07 m/s, the sediment transport rate for sand having a mean diameter of about 0.4 mm is about 20 kg/day/metre width of channel. This corresponds to about 2700 ppm in the sluice channel and a sediment concentration of 700 ppm in the approaching flow of 27 m^3/s, which is acceptable.

Calculations and model studies for flood conditions and for the whole range of sediment sizes are necessary for a thorough check on performance.

APPENDIX 4 : MODEL STUDIES

Model studies of intakes, particularly of major structures, are often necessary because the flow pattern and its interaction with sediment is complicated and defies theoretical treatment; transfer of knowledge and experience from one case study to another is, of course, very valuable but frequently not sufficient, because of special features of design and its interaction with the river banks and bed.

Mathematical models are not yet able to simulate the complex local flow patterns associated with intakes, and therefore, modelling is almost exclusively carried out with scaled physical models.

A4.1 Objectives of Model Tests

The type of model to be used will be primarily determined by the objectives of the model tests. These may be:

(a) Improvement of the overall or detail performance of the intake structure in relation to minimising head losses, avoiding separation, stagnancy, vortex formation, etc..

(b) Improvement of intake design in relation to sediment intrusion, design of sediment exclusion devices and settling basins, design of intakes in relation to the immediate vicinity of the intake, etc..

(c) Interaction between conceptual design and siting of the intake and the river morphology, sediment, ice and floating debris transport, river flow and river training works.

A4.2 Types of Models

In principle three types of models may be considered in isolation, but more frequently jointly:

(i) two or/and three dimensional undistorted scale models of intake structures and of their details (A4.1(a), above).

(ii) three dimensional models of intakes with <u>part</u> of the adjacent river bank and bed; these models frequently do not reproduce the whole river width and extend only a relatively short distance upstream of the intake (near-flow-field modelling). The river is often modelled with a fixed bed but with moveable material added to represent areas where scour and deposition are likely. The models are almost invariably undistorted but occasionally there is a small distortion between vertical and horizontal scales (A4.1(b), above).

(iii) three dimensional moveable bed models where a substantial river reach upstream of the intake as well as the intake itself are modelled; these are often distorted models with only the main features of the intake reproduced (A4.1(c), above).

The first two categories of models may be designed as hydraulic models with water as fluid - this is the normal procedure - but in some instances aero-dynamic modelling, with air flowing in a pressurised system and the water surface reproduced by a smooth fixed or adjustable surface, may be advantageous.

A4.3 Data Required

The information required for model studies of intakes can be classified as follows:

(a) design parameters of the intake itself;

(b) proposed operation of the intake (water levels; discharges, etc.);

(c) the maximum amount of hydrological information which is available on the river on which the intake is to be situated (low, high flows, flow duration curves, etc.). This information is essential particularly for models in A4.2(ii) and (iii) but clearly the levels and velocities in the river have a major effect on intake operating conditions. If the intake is to be situated in an estuary, the wave climate is also necessary;

(d) as much sedimentological information as possible (sediment size distribution, bed and suspended sediment load transport, etc.). This information is essential for models in A4.2(ii) and (iii) particularly, and is also valuable for models in A4.2(i) so that local flow patterns and velocities from the model can be applied to the knowledge of sediment transport data and some judgement made as to the effects of the intake on the sediment regime and the effects of sediment on the intake performance, even if sediment motion is not modelled. Bed levels and bank materials are also required;

(e) information on river operations, ongoing or planned (dredging, navigation, etc.);

(f) winter regime, ice formation and ice floes movement, etc.;

A4.4 Scaling Laws

The selection of scaling laws presents the most difficult problem with regard to model studies of intakes.

The general approach to model design is the same as for modelling on scale models the performance of hydraulic structures (models A4.2(i), above) and/or for models of rivers and open channels (models A4.2(iii), above). The main difficulties in modelling arise from special problems related to intakes e.g. vortex formation and in the case of models combining the structure and river approach (models A4.2(ii), above). The scaling problems related to vortices are complex and not fully resolved as gravity, viscous forces and surface tension are all involved (resulting in incompatible scaling equations according to Froude, Reynolds and Weber laws). This means that often somewhat larger than usual safety factors have to be chosen when extrapolating results from small model tests obtained by the use of Froude scaling laws neglecting effects of viscosity and surface tension.

The modelling of intakes with part of the adjacent flow field (A4.2(ii)) is difficult because a compromise has to be reached between the Froude scaling laws applicable to the nondistorted models of the structure and the laws governing the design of open channel flow models which need not always follow the Froude scaling laws, and/or which frequently would lead to distorted models. The outcome - apart from conventional moveable bed models - could be a nondistorted or only slightly distorted model with possibly a fixed bed where the sediment behaviour is simulated by feeding sediment (possibly lightweight material) into the model and using it as a 'tracer'. Aerodynamic models are another possibility. The choice of the model should be made only after a careful appraisal of advantages and disadvantages of the various approaches.

Where modelling sediment movement in the main water course and the performance of the sediment controlling features of the intake are requirements of the model study the aim is to select scaling laws to ensure that hydraulic and sediment transporting and depositing conditions in the model, correctly reproduce those conditions in the prototype. Many researchers have produced theories for selection of scaling laws and some of these are described in Refs. A4.1 and A4.2. Different criteria apply to suspended load and bed load.

With respect to velocities, flow patterns, etc., standard Froude scaling criteria are applicable and it is these parameters in the intake which govern the deposition and movement of sediment in the structure. If a material is selected for a mobile-bed model or as a tracer in a fixed bed model which can be proved to represent transport of sediment in the main water course, then it is reasonable to assume that this material will react to the intake structure and its sediment controlling features in the same way as the prototype structure will.

Movement of material in the main water course of the model can be checked by comparison with the prototype and if the intake structure is already in existence proving of the model can be even more accurate. The model material properties (size, density, shape, etc.) can be changed during proving and in the case of distorted mobile-bed models, scales and slopes can also be altered. This proving operation may be very expensive and time consuming and so the maximum amount of prototype data (A4.3) at the outset of model design is essential to save time at the proving stage. A thoroughly proven mobile-bed model is a very valuable design tool.

The advantages of using a light weight tracer material in a fixed-bed, undistorted model are that less cost and time are required for design and testing the model. If verification is possible and considered desirable, the single possible adjustment is to change the sediment material until the investigator is satisfied that model results show good agreement with prototype data. The investigator must be confident of his ability to interpret the results of this type of model to provide prototype information.

It is important for the intake designer to discuss with the model investigator the requirements of the model study (number of alternatives to be investigated, etc.) so that the most appropriate model type can be selected and necessary data collected to enable the model investigator to evaluate the costs, time-scales and limitations of the model study.

Modelling assists the designer to perfect his ideas and the more designs that are proved and improved by the use of models, however simple, the better will be designs in the future.

A4.5 Examples of Intake Model Studies

(1) The design of loose boundary physical models of rivers in equilibrium is described in Ref. A4.3 as applied to an intake structure on a river. The selection of model scales developed at HRS is described and a model study of an intake to an irrigation project on the Sabi river in Zimbabwe is described. This was a distorted (6:1) model with a mobile-bed. The study enabled the optimum intake geometry to be selected and the operation of the gates best suited to controlling sediment at the intake during seasonal changes to be recommended.

(2) The model study of the Ambergate River Intake is described in Ref. A4.4.

This model study was undertaken at BHRA to investigate the effects of increasing the abstraction from the River Derwent (England) on head loss and silt movement and deposition. The increase in abstraction was from 91 $M\ell d^{-1}$ to 318 $M\,\ell\,d^{-1}$.

A fixed-bed model was constructed to represent 400 m of the river at a linear, undistorted scale of 1:20. The scaling criteria were established from the Froude law. Diakon MG102, an acrylic polymer, of particle size D = 600 µm and specific gravity of 1.19, was added to the model to represent bed load. Each sedimentation test was run for a period of one hour, since equilibrium was established in this time, for a combination of river flow rates and abstraction rates.

The only data available concerning the river sediment was that it was a fine sand (D_{50} = 100 μm). Calculations were made on this basis using Shields and White theory and it was decided that Diakon was probably the most suitable of available materials. The clients were satisfied that the sediment deposition on the inside of the bend and across the intake mouth were representative of prototype observations.

(3) A mobile-bed model was used, as described in Ref. A4.5 to determine the required sediment intake and corresponding intake layout at a large diversion canal from a river. In this case a limited quantity of sediment was to be diverted into the canal to ensure the canal remained stable. The design of the Jonglei Canal in the Sudan is discussed in the reference.

(4) A hydraulic model was used to compare three water intake lay-outs designed to avoid bed-load transport into the intake (Ref. A4.6). The design of the model, into which sediment was fed at its upstream end and collected and analysed at its downstream end, and the tests on the three intake designs with their sediment excluding devices are described.

(5) The model studies referred to in Chapter 8 example (1) (Ref. 8.1) used a fixed-bed model with polystyrene (D_{50} = 0.4 mm) to simulate the distribution of suspended sand over the vertical. A Rouse similitude criterion: ratio of particle fall velocity to shear velocity was used as the model scale law.

The use of model studies has been recommended throughout this guide and most of the examples cited in earlier chapters have used some form of physical model - the studies are described in the references given at the end of the relevant chapters.

References

A4.1 Avery, P. "The problems of sedimentation modelling with particular reference to river intake models". Proceedings Conference on Hydraulic Modelling of Civil Engineering Structures; BHRA, University of Warwick, Coventry, 1982.

A4.2 Novak, P. and Cabelka, J. "Models in Hydraulic Engineering - Physical Principles and Design Applications". Pitman, London 1981, 459 pp.

A4.3 White, W.R. "A novel loose boundary model for investigating sedimentation problems at an intake". Proceedings Conference on Hydraulic Modelling of Civil Engineering Structures; BHRA, University of Warwick, Coventry, 1982.

A4.4 Avery, P. "Model Study of Ambergate River Intake". BHRA Report RR 1653, March 1981.

A4.5 Klaassen, G.J. "Sediment Intake to Reduce Degradation in an Offtake Canal". Hydraulic Engineering in Resources Development and Management, Proceedings XVII Congress IAHR, Italy, 1979, Vol 6.

A4.6 Jacobsen, J.G. "A Comparison of Three Water Intake Lay-outs to Avoid Bed-Load Transport into the Intake". Hydraulic Engineering in Resources Development and Management, Proceedings XVII Congress, IAHR, Italy, 1979, Vol 6.